T0213252

Antimatter

Beatriz Gato-Rivera

Antimatter

What It Is and Why It's Important in Physics and Everyday Life

 Springer

Beatriz Gato-Rivera 🅾
Instituto de Física Fundamental (IFF), Consejo Superior de
Investigaciones Científicas (CSIC)
Madrid, Spain

ISBN 978-3-030-67790-9   ISBN 978-3-030-67791-6   (eBook)
https://doi.org/10.1007/978-3-030-67791-6

Cover image: Annihilation of an antiproton against an atom of gas neon. Experiment PS-179 in the LEAR
machine at CERN, 1984. © CERN

This Springer imprint is published by the registered company Springer Nature Switzerland AG
The registered company address is: Gewerbestrasse 11, 6330 Cham, Switzerland

*To the memory of Carl Anderson and Dmitri Skobeltzyn,
and to my husband, Bert Schellekens, for his constant support*

Preface

This book is a substantially extended and updated version of the Spanish book *Antimateria*, published in October 2018 by the Editorial CSIC/Los Libros de la Catarata. That book was, in turn, a natural extension of several outreach talks I gave in the CSIC (Spanish National Research Council), in Madrid, on the subject of antimatter, from 2015 until 2017.

The present English version contains additional chapters due to the extensive enlargement of the existing material that led to its rearrangement. One chapter is however written completely anew. It describes the main features of antimatter, dark matter and dark energy, contrasting them with each other, in order to provide a global view of the 'exotic side' of the Universe. I decided to add that chapter to shed some light on these topics since they seem to create some confusion. In fact, antimatter is often confused with dark matter, which, in turn, is also confused with dark energy.

Madrid, Spain
November 2020

Beatriz Gato-Rivera

Acknowledgements

First of all, I am deeply indebted to my colleague and husband Bert Schellekens (Nikhef, Amsterdam) for his continuous support, reading every single line of the manuscript and helping drawing many of the illustrations. His thoughtful comments have contributed to substantial improvements in the redaction of this book.

I am also really grateful to many colleagues for sharing with me very useful information. From cosmologists Emilio Elizalde (ICE-CSIC, Barcelona) and Pilar Ruiz Lapuente (IFF-CSIC, Madrid), I have learned very interesting facts about the early history of Cosmology and the acceleration of the Universe, respectively. From cosmic-ray experts, Manuel Aguilar (CIEMAT, Madrid) and Fernando Arqueros (UCM, Madrid), I got updates and advice about the AMS experiment, on board the International Space Station, and about the Pierre Auger Observatory, respectively. Stefan Ulmer (RIKEN, Ulmer Fundamental Symmetries Laboratory, Japan) provided the last updates on the antimatter experiments conducted at CERN. I have also benefit very much from scientific discussions with particle physicists Sergio Pastor Carpi and Juan José Hernández (IFIC-CSIC, Valencia) who in addition helped me finding data and references on various issues. I wish to thank Alexander Dolgov (Novosibirsk State University and University of Ferrara), for detailed information on the current state of the research about antimatter structures in the Universe, and Horst Breuker (CERN) for a guided visit to CERN's Antimatter Factory in September 2018. I also acknowledge Jan Willem van Holten and Robert Fleischer (Nikhef, Amsterdam) for some useful information about cosmic rays and B meson systems, respectively, and Pierre Sikivie (University of Florida) for some remarks on axion detection.

I really appreciate the help and kindness of Angela Lahee and Christian Caron, from Springer, who from the very beginning supported this project. I also would like to thank the Spanish 'Editorial CSIC' and 'Los Libros de la Catarata' for allowing me to use the Spanish book *Antimateria* as basis to elaborate the present book, which is a very much extended and updated version of the former. In this respect, I am also indebted to Alberto Casas (IFT-CSIC, Madrid) and Fernando Barbero (IEM-CSIC, Madrid) for their many comments to the manuscript of that book.

Finally, I am very grateful to the National Institute of Subatomic Physics (Nikhef, Amsterdam), in The Netherlands, where part of this book was written, for hospitality.

Contents

1 Introduction   1
  1.1 Preliminaries   1
  1.2 Some Basic Concepts   4
       1.2.1 The Speed Limit c   4
       1.2.2 Ions and Plasmas   4
       1.2.3 Isotopes   5
       1.2.4 Kelvin Temperature Scale   5
       1.2.5 Powers of 10   6
       1.2.6 Units of Mass and Energy   6
       1.2.7 Times and Distances   7

2 Antimatter Versus Matter   9
  2.1 Atoms and Antiatoms   10
  2.2 Matter-Antimatter Annihilation   11
  2.3 Other Elementary Particles   13
  2.4 Properties of Elementary Particles   14
       2.4.1 Spin and Helicity   14
       2.4.2 Electric Charge and Electromagnetic Interactions   16
       2.4.3 Strong Charge and Strong Interactions   17
       2.4.4 Weak Charge and Weak Interactions   19
       2.4.5 Mass: Inertia and Gravitation   20
  2.5 The Standard Model   24
       2.5.1 Genesis   24
       2.5.2 Elementary Fermions   30

	2.5.3	Elementary Spin 1 Bosons	35
	2.5.4	The Higgs Boson and the Higgs Mechanism	40
2.6	Beyond the Standard Model		44
	2.6.1	Supersymmetric Particles	45
	2.6.2	Kaluza-Klein Particles	47
	2.6.3	Axions	49
	2.6.4	Dark Matter Particles	52

3 Antimatter Versus Dark Matter and Dark Energy 53

3.1	The Expanding Universe		53
3.2	Antimatter, Dark Matter and Dark Energy: Key Features		62
3.3	Composition of the Universe		65
3.4	Dark Matter: Invisible Source of Gravity		67
	3.4.1	The Case for Dark Matter	67
	3.4.2	Dark Matter Candidates	77
	3.4.3	Dark Matter Detection	80
3.5	Dark Energy: The Accelerating Universe		81
	3.5.1	The Accelerated Expansion of the Universe	81
	3.5.2	Dark Energy Candidates	89
	3.5.3	Dark Energy and the Destiny of the Universe	94

4 Discovering Antimatter 97

4.1	The Dirac Equation		98
4.2	The Positron		103
4.3	Antiprotons and Antineutrons		120
4.4	Antinuclei		123
4.5	Antiatoms		124
4.6	Primordial Versus Secondary Antimatter		126
	4.6.1	Primordial Antimatter	127
	4.6.2	Secondary Antimatter	128

5 Cosmic Rays 137

5.1	The Cosmic Ray Pioneers		137
5.2	Cascades of Cosmic Rays		141
5.3	Ultra-High Energy Cosmic Rays		147
5.4	The Pierre Auger Observatory		150
5.5	The GZK Limit		152
5.6	Detectors in Space		152
	5.6.1	Balloon-Borne Experiments	153
	5.6.2	Experiments on Board Satellites	155
	5.6.3	The AMS-02 Experiment	158

6   Particle Physics Accelerators   167
  6.1   Generalities   167
  6.2   High Energy Collisions   168
        6.2.1   Electron–Positron Collisions   169
        6.2.2   Hadron Collisions   170
  6.3   Brief History of Particle Physics Accelerators   171
  6.4   CERN   182
  6.5   Beyond the LHC   192

7   Matter–Antimatter Asymmetry   193
  7.1   Matter–Antimatter Asymmetry in Astrophysics   193
        7.1.1   Neutrinos or Antineutrinos: The Key Factor   194
        7.1.2   Astronomical Observations: DGRB and CMB   199
        7.1.3   Antistars and Antigalaxies?   202
  7.2   Matter–Antimatter Asymmetry in the Standard Model   204
        7.2.1   Conserved Quantities: q, L and B 205
        7.2.2   Charge Conjugation C 206
        7.2.3   Parity Transformation P 206
        7.2.4 CP Symmetry   208
        7.2.5   Time Reversal T 214
        7.2.6 CPT Symmetry   215
        7.2.7   Particle-Antiparticle Oscillations of Neutral
                Mesons   216
  7.3   Baryogenesis   217
        7.3.1   The Great Annihilation   218
        7.3.2   Baryogenesis and the Sakharov Conditions   220
  7.4   Leptogenesis   224
  7.5   Ettore Majorana and His Fermions   226

8   Experiments with Antiatoms   231
  8.1   Preliminaries   231
  8.2   The Experiments   234
        8.2.1   ATHENA and ATRAP   234
        8.2.2   ALPHA   236
        8.2.3   ASACUSA   239
        8.2.4   BASE   242
        8.2.5   AEGIS   243
        8.2.6   ELENA   248
        8.2.7   GBAR   250
        8.2.8   FAIR   251

9   Medical and Technological Applications of Antimatter   253
   9.1   Medical Applications   254
   9.2   Technological Applications   257
   9.3   Antimatter as an Energy Resource?   261

Appendix A: Atomic Spectroscopy   265

Appendix B: The Myth of Skobeltzyn and the Positron Tracks   269

Epilogue   287

Further Reading   291

About the Author

Beatriz Gato Rivera also known as B. Gato, is a Staff Scientist at the 'Instituto de Física Fundamental' (IFF, Madrid, Spain) of the CSIC (Spanish National Research Council). After finishing her Ph.D. on the topic of Supergravity, she spent three years at the Massachusetts Institute of Technology (MIT, Cambridge, USA) and another three years at CERN (Switzerland), working on String Theory. She has been the Principal Investigator of four research projects. In addition, at the IFF she was the head of the department of 'Particle Physics and Cosmology' for four years, and coordinator of the 'Mathematical Physics' group from 2008 until 2019. Her scientific interests are centered on Mathematical Physics, Particle Physics, Gravitation, and Foundations of Quantum Mechanics. In the last ten years, she has also been involved in outreach activities, especially public talks, and in 2018 she published the book *Antimateria* in the CSIC's collection '¿Que sabemos de?'. This was the basis for the much extended and updated English version now available as the present volume *Antimatter*.

1

Introduction

1.1 Preliminaries

Antimatter is one of the most fascinating aspects of Particle Physics. When one reads or hears the word antimatter, whether in the media, in the cinema, or in a novel, the first thing that comes to mind is: but what are they talking about? In fact, its very name bears some scent of science fiction in it, of something out of this world. This impression is misleading, however, since we actually live surrounded by antimatter and by the products of its annihilation against matter. For one thing, we are immersed in a constant shower of thousands of particles, from matter as well as antimatter, reaching incessantly the surface of the Earth in all directions. They come from the upper layers of the atmosphere, where they are produced by the impact of the cosmic rays against the atomic nuclei of the molecules present there. Some of these particles can even penetrate our homes and buildings traversing everything they encounter in their path, including ourselves.

Stars themselves are an important source of antimatter since it is produced copiously in the plasma of their nuclear furnaces in the form of antielectrons, the so-called positrons. These annihilate rapidly with the electrons in the plasma providing part of the light and heat emitted by the stars. In the case of the Sun, about 10% of the visible light that shines on us these days originated from the electron-positron annihilations that took place within it several hundred thousand years ago. Besides that, some natural radioactive substances, such as Potassium-40, also emit positrons. This makes it possible for a banana to release 15 positrons every 24 h, approximately, from the radioactive nuclei of those atoms. Finally, it should be noted that antimatter is

© Springer Nature Switzerland AG 2021
B. Gato-Rivera, *Antimatter*,
https://doi.org/10.1007/978-3-030-67791-6_1

widely used in our society; in medicine as well as in cutting-edge technology, where it has many applications. As a matter of fact, positrons are the essential ingredient of PET imaging techniques carried out in hospitals all over the world.

The purpose of this book is to explain what antimatter is and many other issues related to it. We will see that it is the reverse of the ordinary matter, but to fully understand this idea one must descend into the realm of elementary particles; where for every kind of existing particle there is another one with opposite properties. Indeed, it is only a matter of convention which ones we call matter particles and which antimatter particles, or antiparticles for short. For this reason, in Chap. 2, after introducing atoms and antiatoms and the subject of matter-antimatter annihilation, we provide an introduction to Particle Physics. We start with the description of the subatomic particles and antiparticles, and the forces between them, with a detailed description of their main properties. Then we review the basic concepts on which the Standard Model is built, as well as some aspects beyond this model.

Chapter 3 explains what dark matter and dark energy are, comparing and contrasting them with antimatter, providing a fairly good introduction to these two subjects. It also describes in some detail the discovery of the expansion of the Universe and the further discovery of its acceleration. This chapter is not only useful on its own but is also included because many people confuse antimatter with the *invisible* components of the Universe (dubbed "dark" for historical reasons).

Chapter 4 reviews the major landmarks on the discovery of antimatter particles, from elementary particles all the way until antiatoms. In particular, the discovery of the first one, the positron, that changed physics forever, is described in very much detail. These findings took place first in cosmic rays and subsequently in particle accelerators, where they were artificially produced. The last section discusses primordial versus secondary antimatter, which is a crucial distinction since the former could have given rise to antimatter structures in the Universe, such as antistars, if it were to exist. Chapters 5 and 6 provide quite complete introductions to the subjects of cosmic rays and particle accelerators, which are the main sources of antimatter accessible to us, apart from some natural radioactive substances.

Chapter 7 deals with the by now famous problem of the matter-antimatter asymmetry in the Universe: *Why is there so much matter compared to antimatter in the Universe?* which is one of the most surprising enigmas in Astrophysics, Particle Physics and Cosmology. We discuss its main aspects in Astrophysics as well as in the Standard Model of Particle Physics, where use is made of the so-called Sakharov conditions as the guiding principle to shed light on the problem of the primordial baryogenesis (the creation of protons and neutrons). The chapter includes a brief account of the possibility of baryogenesis via a special type of neutrinos—leptogenesis—and also discusses some research on primordial antimatter; in particular, the possibility that it had given rise to large structures like antistars and antigalaxies. Finally, it also addresses the enigma of the Italian physicist Ettore Majorana and his fermions.

Chapter 8 describes mainly the experiments performed at the CERN Antimatter Factory to create and analyze antihydrogen atoms with the purpose of comparing their properties with those of the ordinary hydrogen. Chapter 9 addresses the medical and technological applications of antimatter: its use in hospitals to perform the Positron Emission Tomography, known as PET scanner, as well as its utilization for a multitude of research issues in Materials Science and Technology. It also clarifies why it is not feasible to use matter-antimatter annihilation as energy supply to cover the daily needs in our homes and factories, despite the fact that it represents the most energetic process that exists (a thousand times more efficient than nuclear energy).

Appendix A introduces Atomic Spectroscopy, which is an essential tool for both, the study of stars and galaxies in the Universe and the study of atoms and antiatoms in the laboratories. Appendix B debunks a myth about the Russian physicist Dmitri Skobeltzyn and the discovery of the positron, which began in middle 1950s in Cambridge (UK) and spread especially among British scientists.

Finally, the Epilogue discusses some prospects for next years with respect to antimatter research, and also contains a short science-fiction tale in order to illustrate the deep similarities between matter and antimatter.

We welcome all readers to embark on this adventure, this journey to the world of antimatter. But before we begin, we shall introduce some terms and concepts that are repeated throughout the text. We explain first the meaning of the *speed limit c* and what *ions, plasmas* and *isotopes* are. Then we introduce the *Kelvin temperature scale* and describe the *powers of 10*. Finally, we present the *units of mass, energy, time and distance* that are used in Particle Physics, also known as High Energy Physics.

1.2   Some Basic Concepts

1.2.1   The Speed Limit c

The speed limit c is the maximum speed with which a body can move towards or away from another one; its value is 299,792 km/s. In addition, massless particles must travel at this speed through the vacuum, where they do not meet other particles to interact with. This amazing result is deduced from the Theory of Special Relativity, which Albert Einstein formulated in 1905, and implies that it is not possible to accelerate subatomic particles, no matter how much energy is provided, to get them to surpass this maximum speed c. A word of caution should be added, however. Due to the expansion of the Universe, very distant galaxies are typically moving away from our Milky Way and from each other at velocities greater than c, dubbed superluminal velocities. The reason is that space itself grows very fast carrying the galaxies along, and this does not conflict with Special Relativity.

For historical reasons, c is known as the speed of light since it coincides with the speed of electromagnetic waves when they propagate through the vacuum. The reason for this is that electromagnetic waves can be interpreted in terms of massless particles—photons—traveling through space. By contrast, when photons, or equivalently electromagnetic waves, propagate through material media, their speed can be much lower, and even zero inside the opaque media that absorb them; this depends on the characteristics of the material and also on the frequency of the waves. For example, ordinary walls are opaque for the electromagnetic frequencies of visible light, but not for the frequencies corresponding to radio, TV, cell phones, etc. which is why these waves pass through the walls without much difficulty.

1.2.2   Ions and Plasmas

Atoms are composed of a central nucleus, consisting of protons and neutrons, and an outer shell formed by electrons, in equal numbers as protons. In normal conditions atoms are electrically neutral; that is, they have no electric charge because the positive charges of the protons (+1 per proton) are compensated by the negative charges of the electrons (−1 per electron). However, for different reasons atoms can gain or lose electrons. For example, a very energetic particle coming from outside Earth can collide with an atom's electron and pull it out of its orbit around the nucleus, which happens very often in our atmosphere. In these circumstances, the atoms cease to be electrically neutral and are called ionized atoms, or ions. If the ion has excess electrons, it is a negative ion, and otherwise it is a positive ion.

But it can also happen, due to high temperatures or strong electromagnetic fields, that stripping away electrons orbiting the atomic nuclei does not bring them very far apart. Then the ions and the unbound electrons behave like an electrically neutral gas called plasma with (almost) balanced positive and negative electric charges. Plasma is one of the four fundamental states of matter, the other three being solid, liquid and gas, and represents probably the most abundant form of ordinary matter in the Universe. Examples of partially ionized plasmas are lightning and neon adverts, whereas the interior of stars and their coronas consist essentially of fully ionized plasmas.

1.2.3 Isotopes

The atoms of each element in the Periodic Table are characterized by having a fixed number of protons in the nucleus. This is the so-called atomic number, on which the classification of the elements is based. However, the number of neutrons of a given element is variable and characterizes the different isotopes of it. For example, hydrogen is the simplest and lightest element, since it has a single proton in the nucleus, but it can have zero, one or two neutrons when it forms in nature. Therefore, there exist three natural isotopes of hydrogen, although only two are stable, and there are some other isotopes that have been synthesized in laboratories. The most abundant, the ordinary hydrogen ^{1}H, has no neutron in the nucleus; the second isotope, deuterium—denoted as ^{2}H or D—has one neutron; and the third isotope, tritium ^{3}H, has two neutrons, is radioactive and has a half-life of 12.3 years (the time it takes for half of any quantity to decay). The left superscript on the element's symbol indicates the atomic mass number of the isotope, which is the number of protons plus the number of neutrons. The next element, helium, has two protons in the nucleus. It has several isotopes, but only two of them are stable: ^{3}He and ^{4}He, which is the ordinary helium, much more abundant than the first.

1.2.4 Kelvin Temperature Scale

The Kelvin temperature scale, used mainly by scientists, was proposed in 1848 by the British physicist William Thomson, best known as Lord Kelvin. It is an absolute temperature scale since it has an absolute zero below which temperatures do not exist. The reason is that temperature is a measure of energy, and zero Kelvin, 0 K, is the temperature at which atoms and molecules are at their lowest possible energy, the so-called zero-point energy. This lowest energy is non-zero, however, because of quantum fluctuations. The steps in the Kelvin scale - the Kelvin degrees (K) – are of the same size as those of the Celsius scale - the Celsius degrees (°C)–and the correspondence between the two scales is: 0 °C = 273.15 K and 0 K = −273.15 °C, where 0 °C is defined as the freezing temperature of water (and 100 °C as its boiling temperature) at sea-level atmospheric pressure. Although the Celsius scale is used in most of the world, in the USA one uses the Fahrenheit scale, with 180 degrees (denoted °F) between the freezing point of the water (32 °F) and the boiling point (212 °F). In this scale, the absolute zero of temperature corresponds to: 0 K = −459.67 °F.

1.2.5 Powers of 10

The powers of 10, which we write as 10^N, have a very simple meaning. If the power N is a positive number, it indicates the number of zeros to be added after the 1. For example, one thousand is expressed as $10^3 = 1000$; one million as 10^6; one billion as 10^9 and one trillion as 10^{12}. Conversely, if the power is negative, we can write 10^{-N} with N positive, and then the N zeroes go before the 1, which occupies the N-th decimal position. Thus, one thousandth is expressed as $10^{-3} = 0.001$; one millionth (also called micro) as 10^{-6}; one billionth as 10^{-9} and one trillionth as 10^{-12}.

1.2.6 Units of Mass and Energy

The units of mass and energy used in Particle Physics are the same as they are related through the Einstein's mass-energy conversion formula $E = mc^2$. These units are based on the electronvolt, eV, which is the energy that an electron acquires when it is exposed to an electric potential of one volt. The multiples of the eV most commonly used are:

$$1 \text{ keV} = 10^3 \text{ eV}, \ 1 \text{ MeV} = 10^6 \text{ eV}, \ 1 \text{ GeV} = 10^9 \text{ eV},$$

where k stands for kilo, M for Mega, and G for Giga. The electron mass is 511 keV/c^2, that is 0.511 MeV/c^2. This means that the mass of one electron can be transformed into photons with a total energy of 511 keV in suitable processes, such as annihilation with one antielectron. The proton mass is 938 MeV/c^2, hence 1836 times bigger than the electron mass. In kilograms, 1 MeV/c^2 is equivalent to 1.78×10^{-30} kg.

1.2.7   Times and Distances

Unlike the energies, in Particle Physics distances and times are very small in the processes and reactions among the particles. Thus, for distances and times one uses submultiples of the meter and the second, respectively, mainly:

the microsecond, $1\,\mu s = 10^{-6}\,s$ the micrometer, $1\,\mu m = 10^{-6}\,m$
the nanosecond, $1\,ns = 10^{-9}\,s$ the nanometer, $1\,nm = 10^{-9}\,m$
the picosecond, $1\,ps = 10^{-12}\,s$ the picometer, $1\,pm = 10^{-12}\,m$
the femtosecond, $1\,fs = 10^{-15}\,s$ the femtometer, $1\,fm = 10^{-15}\,m$

2

Antimatter Versus Matter

Antimatter can be considered as the reverse of matter. In a broad sense, it is analogous to its mirror image. As we all know from our own experience, when we look at ourselves in a mirror, the face we see is not exactly ours but has the right and left sides interchanged. Similarly, antimatter particles have opposite properties with respect to matter particles. This refers to all properties that admit opposite values, such as electric charge; but there also exist properties that do not admit opposite values, like the mass, and these are identical for the particles and their antiparticles. For example, as we will see later, quarks and their antiquarks have the same mass, the same spin and the same mean lifetime, but opposite values of the strong charge, the electric charge, the weak charge and the baryon number. Analogously, the electron and its antiparticle, the positron, have the same mass, the same spin and the same mean lifetime, but opposite values of the electric charge, the weak charge and the lepton number. Curiously, the positron is the only antiparticle bearing its own name; the other antiparticles are named like the ordinary particles but with the prefix *anti*. Indeed, all elementary particles have antiparticle partners, although a few of them are actually their own antiparticles, as in the case of the photon—the particle of light—and the Higgs boson.

In this chapter we will explain in some detail what matter particles and antimatter particles are, which are their properties and which forces and interactions they experience. We will also present the Standard Model of Particle Physics, and discuss briefly some proposals "Beyond the Standard Model" to address some problems that remain to be solved. To start, we will have a first encounter with atoms and antiatoms, as well as with the very remarkable issue of the matter–antimatter annihilation.

© Springer Nature Switzerland AG 2021
B. Gato-Rivera, *Antimatter*,
https://doi.org/10.1007/978-3-030-67791-6_2

2.1   Atoms and Antiatoms

As the Greek philosopher Leucippus (5th Cent. B.C.)[1] and his pupil Democritus (460 B.C.–370 B.C.) already anticipated, the entire material world in which we are immersed, and which forms our bodies, is made up of atoms. These bind together to form molecules, crystals, and other structures that build all the solids, liquids and gases that we perceive. However, unlike the atomic model of Leucippus and Democritus, in which these corpuscles were elementary, immutable, and indivisible, real atoms have a structure. They consist of a nucleus and a shell, and can be broken down and divided into their constituent subatomic particles: the nucleons—*protons* and *neutrons*—in the atomic nuclei and the *electrons* in the shells, orbiting the nuclei due to the electrical attraction between their negative charges and the positive charges of the protons.

Moreover, it turns out that protons and neutrons are not elementary either, but are composed of particles that are believed to be elementary: the *quarks*. These are bound together through the strong interactions, as we will discuss later. There are six types of quarks although only two of them are constituents of protons and neutrons. The latter are called quarks u and d, from up and down. The proton p is composed of two quarks u and one quark d, while the neutron n is constituted by one quark u and two quarks d:

$$p = (u, u, d), \quad n = (u, d, d). \tag{2.1}$$

The electron e^- also seems to be elementary. This actually means that Particle Physics experiments are not able to detect any structure, at present, neither in the quarks nor in the electrons. But if such structures existed and were detected in future experiments, these particles would no longer be considered elementary.

The antimatter atoms, on the other hand, are formed with the antiparticles of the particles that make up matter; i.e. with *antiprotons*, *antineutrons* and *positrons*. The electric charge of the atomic antinuclei is negative because the antiprotons have the opposite electric charge to that of protons; and orbiting the antinuclei one finds the positrons e^+ with a positive electric charge opposite to that of electrons. The antiprotons $\overline{p}$ and antineutrons $\overline{n}$ are composed of the antiquarks $\overline{u}$ and $\overline{d}$, in a similar way as the composition of protons and neutrons:

$$\overline{p} = (\overline{u}, \overline{u}, \overline{d}), \quad \overline{n} = (\overline{u}, \overline{d}, \overline{d}), \tag{2.2}$$

---

[1]The dates of Leucippus's birth and death are not recorded. Aristotle and Theophrastos explicitly credited him as the originator of atomism.

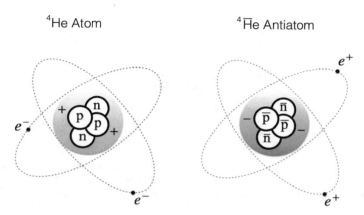

Fig. 2.1 Sketch of a Helium-4 atom and an Antihelium-4 antiatom. The nucleus of the atom, composed by two protons and two neutrons, has positive electric charge while the nucleus of the antiatom, composed by two antiprotons and two antineutrons, is negatively charged. These nuclei are named alpha particles and antialpha particles, respectively

where the bar over the symbol of the particles indicates that they are antiparticles.

In Fig. 2.1 we see a sketch of the helium atom ^{4}He, its most abundant isotope, and of its antimatter counterpart, the antihelium atom $^4\overline{\text{He}}$. They are not to scale since the distance between the atomic shells and the nuclei is actually about 100,000 times the size of the latter. The nucleus of the ^{4}He atom, composed by two protons and two neutrons, results to be a very stable—hard to break—configuration. It is called *alpha* particle, denoted as α, because the rays with the same name, discovered at the end of the nineteenth century in natural radioactivity, consist precisely of these particles when they are emitted by very massive unstable nuclei. Similarly, the nucleus of the $^4\overline{\text{He}}$ antiatom, formed by two antiprotons and two antineutrons, which is just as stable, is called *antialfa* particle and is denoted as $\overline{\alpha}$. Now, unlike α particles, which are very abundant in the Universe since its very beginning, $\overline{\alpha}$ antiparticles have never been detected in nature so far, although their discovery in the cosmic rays could be approaching, as will be discussed in Chaps. 5 and 7.

2.2   Matter-Antimatter Annihilation

Perhaps the most distinctive feature of antimatter is that when it comes into contact with matter they annihilate each other producing a large amount of radiation. Indeed, if a sufficient amount of antimatter could be stored, even

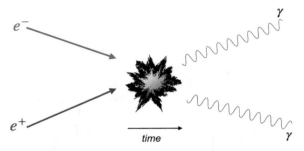

Fig. 2.2 Electron–positron annihilation producing two photons. The masses of the electron e^- and positron e^+ (511 keV/c^2) are totally transformed into the energies of the photons γ (511 keV), according to the mass-energy equivalence $E = mc^2$. This process can also occur in the opposite direction, and then it is called $e^+ e^-$ pair production

quite small compared to the amount of fuel present in nuclear weapons, a bomb could be built with an energy about a thousand times higher than that provided by the nuclear fusion of an equivalent mass. Suffice it to say that a single gram of antimatter would produce, upon contact with matter, a deflagration equivalent to more than twice the atomic bomb that struck Hiroshima in 1945. Fortunately, this enterprise is not easy, as we will see in Chap. 9.

When matter is annihilated with antimatter, each individual matter particle is annihilated with its corresponding antimatter particle—its antiparticle—resulting mainly in photons, denoted as γ, which are the smallest packages or quanta that constitute the electromagnetic radiation. Therefore, when an atom makes contact with an antiatom, each electron e^- is annihilated with a positron e^+, each quark u is annihilated with an antiquark $\bar{u}$ and each quark d is annihilated with an antiquark $\bar{d}$. As a result, the quarks of protons and neutrons are annihilated against the antiquarks of both antiprotons and antineutrons.[2]

In Fig. 2.2 one can see an illustration of an electron–positron annihilation resulting in two photons. As the photons γ are massless, the masses of the electron e^- and the positron e^+ (511 keV/c^2 each) disappear altogether, being transformed into the energies of the photons (511 keV each), as follows from Einstein's formula of mass-energy conversion: $E = mc^2$, where c is the speed limit. This is a fairly good description when the encounters between

---

[2]Although the simplest annihilations of protons with antiprotons and neutrons with antineutrons giving just photons do exist, they are not the most likely annihilation processes because nucleons are very complicated systems. The dominant processes for nucleon-antinucleon interactions - $\bar{p}\,p$, $\bar{p}\,n$, $\bar{n}\,p$ and $\bar{n}\,n$ - occur through pion production, as will be explained in Sect. 2.4.3.

matter and antimatter particles take place at small speeds, that is at low energies. But reality becomes more complex when the particles collide at high energies moving near the speed limit[3] c. This is so because the energy of the collision is invested in creating also massive particles, in addition to photons, making use of the mass-energy conversion, as is usually the case in collisions where new particles are produced which are different from the original ones.

2.3 Other Elementary Particles

So far we have mentioned only the elementary particles that compose the atoms: the quarks u and d in the nuclei and the electrons e^- in the atomic shells. However, there exist many other elementary particles that are considered matter although they are not part of the composition of atoms. This circumstance created some sort of philosophical turmoil when Carl Anderson discovered in 1936 the first of these particles, the *muon* μ^-, in the cosmic rays, a particle 207 times more massive than the electron but otherwise identical to it. The problem is that no meaning was seen in its very existence, to the extent that the phrase *Who ordered the muon?* became very popular among physicists.

Apart from the muon μ^-, the other particles to which we refer are: the *tau* particle or *tauon* τ^-, which also has identical properties as electrons and muons but with a larger mass; three types of *neutrinos*: ν_e, ν_μ and ν_τ, mysterious particles in more than one sense, without electric charge and with hardly any mass; and the quarks of type c (*charm*), s (*strange*), t (*top*) and b (*bottom*). Moreover, one has to add to this list the antiparticles of all these particles. For example, the leptons μ^+ and τ^+, also called *positive muon* and *positive tauon*, are the antiparticles of the leptons μ^- and τ^-, where the term *lepton* denotes the matter particles without strong charge (only quarks have it).

Nevertheless, it turns out that, with the exception of neutrinos, these other particles of matter and antimatter decay rapidly. Indeed, the most long-lived ones, the muons, only exist for 2.2×10^{-6} s, so just a few millionths of a second. They do this spontaneously with the assistance of the weak interactions, which is why these particles are said to be unstable. Their fleeting existence arises when they are created by particles colliding with each other.

There are also elementary particles that are not considered matter or antimatter and are responsible for the interactions, i.e. the forces, between the particles. These interactions come in four types: *electromagnetic, strong, weak*

---

[3] See the tutorial in the Introduction for more details about the maximum speed c, customarily called "the speed of light".

and *gravitational*. In Sect. 2.5.4 we will explain in detail the properties of these force-carrying particles, but we already anticipate that electromagnetic interactions consist of an exchange of photons between particles that have electric charge; strong interactions are due to the exchange of *gluons* between particles that have strong charge (also called *color*); and weak interactions, which are responsible for most particle decays, result from the exchange of the *bosons* W^+, W^- and Z^0, which are highly unstable themselves. Curiously, the bosons W^+ and W^- are antiparticles of one another whereas the photons and the bosons Z^0 are antiparticles of themselves. As for gravitation, in Particle Physics it is supposed to be mediated by the exchange of the hypothetical *gravitons*, but the experimental verification of their existence is totally out of reach. Finally, we have the Higgs boson H, which is its own antiparticle and is the mediator of an extremely weak force of extremely short range, which is why it is not counted among the forces.

2.4   Properties of Elementary Particles

We have seen that the only difference between matter and antimatter atoms resides in the elementary particles that compose them, which are known as antiparticles in the case of antimatter. The main properties of elementary particles are spin, helicity, mass, electric charge, strong charge, weak charge, baryon number and lepton number. Two of these properties, the spin and the mass, do not admit opposite values and hence are identical for each particle and its antiparticle, but the other properties can take opposite values and are the ones that differentiate particles from their antiparticles, as we pointed out at the beginning of this chapter. In the following paragraphs we will review these properties, except for the baryon and lepton numbers, which are also called baryonic and leptonic charges,[4] and will be introduced in Chap. 7.

2.4.1   Spin and Helicity

Spin is a quantum property similar to an intrinsic angular momentum corresponding to an internal rotation. As it is usual with quantum properties, we lack intuition for the spin, i.e. it is not a rotation like those we observe in our daily life described by Classical Physics. It can take integer values (in appropriate units), such as 0,1,2, in which case the particles are called *bosons*; or

---

[4]The terms baryonic and leptonic "charges" were, and still are, widely used by particle physicists and cosmologists from the previous Soviet Union.

it can take half-integer values, such as 1/2 or 3/2, in which case the particles are called *fermions*. Examples of the latter are all matter and antimatter elementary particles, with spin ½, whereas all the elementary particles which mediate the interactions are spin 1 bosons.

The *helicity* is the sense of the spin with respect to the direction of motion of the particle, so it can be right-handed (clockwise) or left-handed (anticlockwise), taking opposite directions for particles and antiparticles. This property may seem irrelevant to the readers, who may wonder: *what difference it makes if a particle is rotating in one direction or the other?* However, as we will see later, this property is crucial since weak interactions manifest themselves differently depending on the helicity of the particles.

The bosonic or fermionic character of elementary particles is a crucial aspect that has many implications and determines their collective behavior. It also determines the collective behavior of composite particles and atoms because individual spins are combined in such way that an odd number of fermions gives rise to a fermion and an even number to a boson. As a consequence, some atoms are bosons, like ^{4}He, whereas some others are fermions, like ^{3}He. Identical fermions, whether particles or atoms, never share their quantum state, a result known as the *Pauli Exclusion Principle*. As a result, if in a given physical system two seemingly identical fermions have the same energy then they should differ in at least one property. Identical bosons, on the contrary, have the tendency to cluster together and share their quantum state. For this reason, only bosons can form the so-called Bose–Einstein condensates (Fig. 2.3), with a large number of them in the same lowest energy state, or ground state. It should be noted, however, that these condensates are

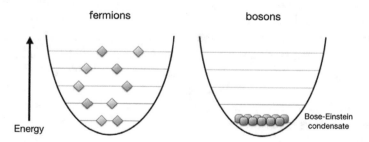

Fig. 2.3 Bosons and fermions have completely different collective behavior. Identical bosons cluster together sharing their quantum state, which is why they can form Bose–Einstein condensates, with a large number of them occupying the state with lowest energy. Identical fermions, on the contrary, never share their quantum state. So, if two seemingly identical fermions have the same energy, then they should differ in at least one property, represented here by the two different colors

only possible provided the temperature is extremely low, near the absolute zero (0 K) to prevent thermal fluctuations to interfere with the process.

To appreciate the importance of the fermionic or bosonic character of the particles, let us note that the stability of atoms, and their own existence as we know them, depend entirely on the fermionic nature of the electrons, with spin ½, which prevents them from descending all simultaneously to the ground state. The reason is that in the Universe it rules, so to speak, the law of minimum effort or minimum energy. Accordingly, if the electrons were bosons, then all of them, distributed in layers and orbitals with increasing energy levels around the nuclei, would fall to the bottom layer, to the ground state orbital. There they would cluster together sharing the same quantum state, so that the other orbitals would disappear de facto and with them the present electronic structure of atoms. But it is this electronic structure what determines the practical totality of the physico-chemical properties of atoms, including the formation of molecules, crystals and other structures, as well as all chemical reactions. In essence, if the electrons were bosons, the material Universe would be very different from the one we know: far more boring, with hardly any diversity and, almost certainly, incapable of harboring life.

2.4.2 Electric Charge and Electromagnetic Interactions

The *electric charge* is carried by all particles that experience electromagnetic forces. These are all the matter particles and their antiparticles, with the exception of neutrinos, as well as the bosons W^+ and W^-. The electric charge can be positive or negative, although the assignment of which charges are called positive and which negative is pure convention, since the charge of the electron could have been called positive and the charge of the proton negative. The relevant fact is that electromagnetic forces, which consist of an exchange of photons, can be attractive or repulsive: electric charges with the same sign—like charges—repel each other, while electric charges with opposite signs attract one another. Consequently, by having opposite electric charges, particles and their antiparticles profess a fatal attraction that often costs them their very existence.

To help intuition it is very useful to use the concept of field, introduced by Michael Faraday in the mid-nineteenth century, according to which electric charges create an electrostatic field around them. The properties of this field can be deduced from Coulomb's Law (1785) that describes the electrostatic force between two charges. This field develops a magnetic component, surprisingly, if the charges are moving with respect to the measuring device;

that is, with respect to the observer.[5] Particles without electric charge—neutral particles—not only do not create their own electromagnetic fields but, in addition, they are not sensitive to the electromagnetic fields in which they are immersed. And conversely, neutral particles are invisible to the electric and magnetic fields around them because these fields cannot detect them.

2.4.3 Strong Charge and Strong Interactions

The *strong charge* or *color* is a property of quarks, antiquarks, and gluons, exclusively, as these are the only particles sensitive to the strong forces. These are also known as "strong nuclear forces" because their range of action is very short and are confined within the atomic nuclei. This charge gives rise to purely attractive forces consisting of an exchange of eight types of gluons (the name gluon comes from glue due to their ability to strongly tie the quarks together). Unlike the electric charge, the strong charge comes in three different kinds, which are called colors in analogy with the three primary[6] colors: red, green and blue. Antiquarks have anticolors, which are antired, antigreen and antiblue. Interestingly, also gluons, which are the mediators of these interactions, are equipped with strong charge. This consists of a color-anticolor pair, for example blue-antigreen.

The main function performed by the strong forces is to bind quarks (and antiquarks) together, forming composite particles called *hadrons* ("dense" in Greek). Protons and neutrons are the most relevant hadrons and also receive the additional name of *baryons,* meaning that they are formed by three quarks. Interestingly, it was found that each of these three quarks has to come in a different color for the total to be neutral, what provides an analogy with the fact that the three primary colors combined together result in white. There are also hadrons formed, curiously, by one quark and one antiquark, with opposite colors so that the total strong charge is again neutral. They are named *mesons* and their mean lifetime is very short (2.6×10^{-8} s the longest). There is a whole variety of these mesons and they are classified into groups, or systems, whose members can be electrically charged or neutral. The most relevant mesons for Particle Physics are: *pions* π, which are composed of a quark and an antiquark of the same species that form protons and antiprotons, *kaons* K, and the B and B_s mesons. Quarks can also come in groups of four,

---

[5]In Physics, observers denote detectors in general, whether they are instruments, human beings, animals or even just atoms.

[6]In Physics, the three primary colors are red, green and blue. In Fine Arts, the green is replaced by yellow as primary color. Anyway, the three "colors" of the strong interactions are just labels that bear no relation with the colors of the electromagnetic spectrum that we observe.

five or six, but these compounds have extremely low probabilities of forming. In Fig. 2.4 one can see the simplified internal structure of a few baryons and antibaryons: the proton p, neutron n, lambda Λ_b and their antiparticles; and also of some mesons: the positive pion π^+, the neutral kaon K^0 and the neutral B^0 and B_s^0 mesons, where the superscript 0 indicates that they are electrically neutral. In Chap. 6 we will see that this structure is actually much more complex.

A residual form of the strong force is what holds protons and neutrons together to form atomic nuclei, even though the protons feel electrostatic repulsion towards each other as they are all positively charged. From this fact alone one can deduce that the strong force must be much more intense than the electrostatic repulsion between the protons, since it is able to overcome that repulsion even in its residual form. The carriers of this residual strong force are, curiously, not gluons but pions π, which are exchanged between nucleons. Another aspect of the nucleon pion exchange is that antiproton collisions against atomic nuclei—$\overline{p}\,p$ and $\overline{p}\,n$—produce typically pions, among other particles. The charged pions most often decay creating muons, which in turn decay giving rise to electrons or positrons. This sequence of decays—pions to muons to electrons or positrons—can be observed in Fig. 2.5, that shows the trails left by the annihilation of an antiproton $\overline{p}$ against a nucleus of a neon atom, as recorded in a photographic plate in a streamer chamber at CERN in 1984. The decays of the positive pion π^+ and

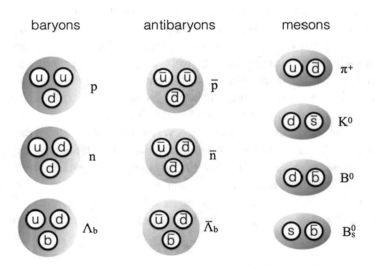

Fig. 2.4 Some examples of hadrons: baryons, antibaryons and mesons. Baryons and antibaryons are fermions, as they are formed by three spin ½ particles, while mesons are formed by two of them and therefore are bosons

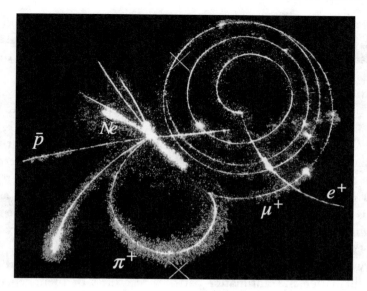

Fig. 2.5 Antiproton annihilation against a nucleus of a neon atom, corresponding to the experiment PS-179 in the LEAR machine at CERN, in 1984. Among the products of the annihilation, a positive pion is created that moves in a spiral downwards, due to the effect of the applied magnetic field, and decays producing an antimuon, which revolves several times until it decays in turn, giving a positron (antielectron). In these decays neutrinos and/or antineutrinos are also emitted, as shown in (2.3), but they cannot be observed since being neutral they leave no trail. *Credit* Courtesy of CERN

the positive muon μ^+ (antimuon) also produce neutrinos and antineutrinos and are as follows:

$$\pi^+ \rightarrow \mu^+ + \nu_\mu \quad \text{and} \quad \mu^+ \rightarrow e^+ + \nu_e + \bar{\nu}_\mu. \tag{2.3}$$

2.4.4 Weak Charge and Weak Interactions

The *weak charge* can be found in all particles sensitive to the weak forces, which also have a very short range, like the strong forces, so that they are also known as "weak nuclear forces". Interestingly, these interactions come in two very different types depending on whether the mediator bosons are the electrically charged ones, W^+ and W^-, or if it is the neutral Z^0 boson. The ratio between the strengths of the weak forces and the electrostatic force at a distance of the proton size (10^{-15} m), and also at distances 100 times shorter, is less than 10^{-4}. Therefore, the weak forces deserve this name because they are several orders of magnitude less intense than the electromagnetic forces.

However, at even shorter distances these two types of forces are unified into a single one, known as the electroweak force. As a result, at those ultra-short distances the weak and the electromagnetic interactions have the same strength.

A distinctive and unique aspect of the forces mediated by the bosons W^+ and W^- is that they are felt by only the matter particles with left-handed helicity and their antimatter partners with right-handed helicity. This implies that the W^+ and W^- bosons do not notice the presence of right-handed matter particles, nor of their left-handed antiparticles. In technical terms, one says that the W^+ and W^- bosons do not have couplings with those particles. Another remarkable aspect of the interactions mediated by the W^+ and W^- bosons is that they change the identity of the particles. Hence, these interactions cannot be described in terms of attraction and repulsion between the particles which interact. We will return to this in Sect. 2.5.4.

The weak interactions mediated by the Z^0 boson, however, are felt by almost all matter and antimatter particles regardless of their helicity. The only exceptions are the right-handed neutrinos and their antiparticles, the left-handed antineutrinos. But it is not known for sure if these particles really exist, since they have never been detected. In addition, in the case that right-handed neutrinos existed they would be practically undetectable because they would interact with the other particles only through gravity, which is an extremely weak interaction, as we will see below. These weak forces mediated by the Z^0 boson can be described in terms of attraction and repulsion between the particles, as they do not change their identity; in this sense they are more intuitive than the interactions mediated by the W^+ and W^- bosons.

A key feature of the weak interactions is that they are responsible for the spontaneous decay of many particles, like the β-decay of the neutron, in which an electron is emitted. The resulting electron radiation was discovered in 1896 by Henri Becquerel in the natural radioactivity of some uranium samples, together with two other radiations of different nature. Shortly afterwards, in 1899, Ernest Rutherford properly identified the properties of all three radiations, which he baptized α-rays, β-rays and γ-rays.

2.4.5   Mass: Inertia and Gravitation

Mass is a multifaceted property of particles and bodies in general. On one side, it gives bodies their *inertia*—resistance against changes in their state of motion—whether they move at a constant speed or are at rest (which is only the particular case of constant zero speed). As the second law of Newton states, $F = m\,a$, for a body of mass m to modify its speed and acquire an acceleration *a*, it is necessary to apply to this body a force *F* proportional to its mass. Therefore, if the mass of an object is small, small forces are needed to move it, or to modify its speed in the case that it is already in motion. And the opposite occurs if the mass of the object is big.

On the other side, mass can be considered like an extremely concentrated energy, like solidified, as Albert Einstein (Fig. 2.6) taught us in 1905 with his world-famous equation of conversion between mass and energy: $E = m c^2$, in the theory of *Special Relativity*. Indeed, the energy released by nuclear weapons is produced through this mass-energy conversion because the final products of the nuclear fission, or fusion, have a smaller mass than the initial products, the difference being converted into the bomb's energy. In the case of the matter–antimatter annihilation, it is the whole mass of the particles put into contact that disappears producing radiation, as shown in Fig. 2.2.

Finally, mass also acts as a gravitational charge of particles and bodies because it creates a gravitational field around them, in the same way that an electric charge creates an electromagnetic field around itself. The force of gravity between any two bodies is described by the Law of Universal

Fig. 2.6 Albert Einstein (1879–1955) in 1904 (left) at the time when he was working at the Swiss Patent Office in Bern (Switzerland), only one year before his "annulus mirabilis" (1905), in which he published four revolutionary articles which included: the Theory of Special Relativity, the most famous equation in history, $E = m c^2$, the photoelectric effect, which lies at the foundation of quantum mechanics, and Brownian motion, which proves that atoms and molecules exist. Ten years later, on November 25, 1915, he presented his theory of gravitation, called Theory of General Relativity. In 1921 (right) he was awarded the Nobel Prize in Physics for his services to theoretical physics, especially for the photoelectric effect. *Credit* The Nobel Foundation

Gravitation, which Isaac Newton published in 1687. Although this law was superseded in 1915 by Albert Einstein's theory of gravitation, known as *Theory of General Relativity*, in practice it is very precise and therefore still broadly used. Nevertheless, Einstein's theory is necessary for GPS devices to make their computations with the greatest precision.

General Relativity

General Relativity is a very abstract theory of gravitation because it does not use the concept of forces between bodies. Instead, it makes use of the *spacetime continuum* and its geometry, and postulates that masses and any other type of energy curve this continuum and they move along the geodesic lines of its geometry (the shortest distances between two points). It is this movement following the geodesic lines what creates the "illusion" of an attraction force between the bodies. Einstein explained his theory in the presentation *"Die Feldgleichungen der Gravitation"* (The Field Equations of Gravitation), on November 25, 1915, in the Royal Prussian Academy of Science, in Berlin (Germany). This presentation was published in the proceedings shortly afterwards. The Einstein equations read:

$$R_{\mu\nu} - \frac{1}{2} R \, g_{\mu\nu} = \frac{8\pi G}{c^4} T_{\mu\nu}, \tag{2.4}$$

where G is Newton's constant and c the speed limit. The subindices μ and ν can take four different values, three for the three space dimensions and one for the time. For this reason (2.4) is not a single equation but a set of equations written in a concise form. On the left hand side, the Ricci curvature tensor $R_{\mu\nu}$ and the scalar curvature R encode the geometry of the spacetime at hand, while, on the right hand side, the stress–energy tensor $T_{\mu\nu}$ encodes the specific matter-energy configuration. There is a famous phrase by John Wheeler, a very prominent physicist from the middle of last century, that describes these equations in a simple way: matter-energy tells spacetime how to curve and spacetime geometry tells matter-energy how to move. Translating this into terms of fields and charges, we see that General Relativity implies that all forms of energy, and not only mass, act as gravitational charges: they are sources of gravitational fields by bending space–time, while at the same time they feel the gravitational fields created by other bodies. For this reason, light, consisting of photons that have energy but no mass, bends appreciably when passing through intense gravitational fields, such as those of our Sun, stars, galaxies, etc. in such a way that this effect is measurable and provides the basis for gravitational lensing.

Compared to the other three forces, gravitation appears to be extremely weak. Indeed, the relation between its intensity and the intensity of the electrostatic force is more than 35 orders of magnitude lower, of the order of 10^{-36}, at a distance of 10^{-15} m. A remarkable difference between these two forces is that gravitation is always attractive and there is only one type of mass, while electromagnetic forces can be attractive or repulsive, due to the existence of two types of opposite electric charges (positive and negative). Another difference, and in fact quite curious, between these two forces is that electromagnetism can be shielded (with a Faraday cage, for example) so that the electromagnetic fields disappear from a region of space, while gravity cannot be shielded; that is, gravitational fields pass through all materials.

As a matter of fact, the only way to cancel gravity, in the sense of not noticing its effects, is the free fall in the given gravitational field, which may seem paradoxical. For this reason, astronauts practice weightlessness in planes that are put briefly in free fall. In this respect, it should be stressed that all bodies moving in a stable orbit around another much larger body are in free fall in its gravitational field. This is the case for all satellites, whether natural or artificial (assuming, of course, that their engines are off). This requires these bodies to reach a certain speed depending on the distance to the larger body. For example, to stay in a stable orbit, the International Space Station, which is about 400 km high, turns around the Earth at the speed of 27,743 km/h, completing one revolution every 92 min. Being in free fall in the Earth's gravitational field, neither the crew nor any object abord feel the gravitational forces—this is why they are floating—in fact they do not even notice that they are moving through outer space (they can see this through the windows, but they do not feel it). In the same way, we here on Earth do not feel that we are moving around the Sun, even though we do it at an average speed as high as 106,200 km/h. This is so because the Earth revolves around it in a stable orbit, and is therefore in free fall with respect to the Sun's gravitational field.

Returning to the subject of the similarities and differences between particles and antiparticles, it should be noted that the mass, in principle, is identical for particles and their antiparticles, as stated earlier. However, although General Relativity implies that gravitation between all bodies must always be attractive, it is still to be confirmed experimentally whether this force is attractive, or repulsive, between a body of matter and another of antimatter. In our opinion, the very fact that some particles are their own antiparticles makes it clear that particles and antiparticles must behave in exactly the same way with respect to gravity. In Chapter 8 we will see the experiments being carried out at CERN to elucidate this question.

2.5   The Standard Model

2.5.1   Genesis

The known elementary particles, together with their properties, constitute the essence of the *Standard Model of Particle Physics*. This model was constructed in several steps, mainly during the 1960s and 1970s, in the framework of Quantum Field Theory, which in turn was supported by the pillars of modern physics: Quantum Mechanics and Special Relativity. Indeed, already in the mid-twenties of the last century, physicists realized that in order to describe the elementary particles they needed to find equations that combined the physics of the minuscule—Quantum Mechanics—with the physical description of processes at high speeds—Special Relativity. This task turned out to be quite difficult, but in 1927 Paul Dirac (Fig. 2.7) succeeded in writing the relativistic equation that bears his name, which describes the electron and any other elementary fermion of spin 1/2. This equation, published in 1928, allowed Dirac to predict the existence of antimatter, as will be discussed in Chap. 4, where we will present an overview of the major discoveries of antiparticles.

Fig. 2.7   Paul Adrien Dirac (1902–1984) when at the Cavendish Laboratory in Cambridge University (UK), around 1930. He wrote down the equation which describes the behavior of relativistic electrons and predicts the existence of positrons. He also predicted the existence of antiprotons and the production of particle-antiparticle pairs from very energetic photons. He is considered one of the most prominent physicists of the twentieth century, and shared the 1933 Nobel Prize in Physics with Erwin Schrödinger. *Credit* Courtesy of Wikimedia Commons

In the following decades, the theories that integrated all the other particles were developed, starting with the theory that describes electrons and their electromagnetic forces mediated by photons, a theory that was named *Quantum Electrodynamics* and is known as QED. One of the most important physicists who built this theory was Richard Feynman (Fig. 2.8), creator of the diagrams that bear his name and are the tool used to calculate the outcome of the processes between elementary particles.

These theoretical developments were also occurring in parallel, in the 1950s and 1960s, with an extensive harvest of experimental findings obtained in particle accelerators, consisting mainly in the discovery of particles sensitive to strong forces, which were termed hadrons. However, so many and so varied were the hadrons that, around 1964, George Zweig and Murray Gell-Mann (Fig. 2.9) proposed, independently, that these were particles composed

Fig. 2.8 Richard Phillips Feynman (1918–1988) in 1965 when he was awarded the Nobel Prize in Physics along with Julian Schwinger and Shinichiro Tomonaga. He is regarded as one of the ten greatest physicists of all time. During his lifetime he became one of the best-known scientists in the world, partly due to his three-volume publication "The Feynman Lectures on Physics" and several popular books he wrote, but also because he was a member of the panel that investigated the disaster of the Space Shuttle Challenger in 1986. *Credit* The Nobel Foundation

Fig. 2.9 Murray Gell-Mann (1929–2019) when he received the 1969 Nobel Prize in Physics for his contributions to the theory of elementary particles and their interactions. He and George Zweig postulated in 1964, independently, the existence of constituents of the hadrons, that Gell-Mann called quarks. He was a particularly talented character in many disciplines, inside as well as outside Physics. *Credit* The Nobel Foundation

of elementary spin 1/2 fermions, which the latter dubbed quarks.[7] Moreover, in 1971 Gell-Mann and Harald Fritzsch proposed the existence of the strong charge, that they named "color" charge.

After many tedious calculations with endless difficulties, in which numerous physicists participated, what was found, in summary, is that the theory that describes elementary particles and their interactions, with the exception of gravitation, has as its fundamental ingredients not particles but *quantum fields* that extend throughout space and permeate the entire Universe. And this is the case not only for matter and antimatter particles but also for the particles mediating the interactions. Namely, in this theory all elementary particles, without exception, are reduced to excitations or perturbations of quantum fields, to wave packets thereof, in a similar way as when waves are produced on the surface of a pond by throwing a stone. To be more

---

[7] Gell-Mann found the term quark in a poem by James Joyce: "Three quarks for Muster Mark!"

precise, there is a different quantum field for each particle species, except that a particle and its antiparticle share the same quantum field. Hence, we have: the field of the photon, the field of the electron and positron, the field of the muon and antimuon, three fields for the three neutrino and antineutrino species, eights fields for the eight types of gluons, the field of the quark and antiquark u and $\bar{u}$, the field of the Higgs boson, etc.

An important remark is that in Particle Physics, or more precisely in Quantum Field Theory, one uses the term *vacuum* to denote the state of minimum energy of all the quantum fields simultaneously. Hence, since the particles are excitations of the quantum fields, the vacuum is the state where there are no particles at all (nor even photons).

This concept of vacuum is a formal, stricter version of the term vacuum commonly used to describe regions of space, of all sizes, that are more or less "empty" rather than devoid of any particles. This applies to devices or containers that have been subjected to pumping procedures to remove all possible matter from their interior, all the way up to large regions of outer space. These "empty" containers, as well as the vast "empty" regions of outer space, contain, however, a multitude of particles. For one thing, it is impossible to block all the neutrinos passing by, that enter everywhere. The same can be said of many photons coming from all directions, especially the thermal ones produced by any matter above the absolute zero of temperature (0 K), like the materials which form the walls of the aforementioned containers or devices. As for gravity, it is impossible to block gravitational fields from any region of space at all, and even less those created by the containers or devices themselves.

The vacuum of Quantum Field Theory, however, is full with all the quantum fields corresponding to the elementary particles and these fields cannot be removed. Not only that, but in addition these fields manifest themselves through, at least, quantum fluctuations and endow the vacuum with energy. As a consequence, the vacuum has a minimum energy different from zero even in the absence of any particles. Furthermore, since 2012, thanks to the discovery of the Higgs boson we know that the corresponding quantum field does exist and manifests itself not only through quantum fluctuations but also generating masses for most elementary particles, including itself.

Observe, moreover, that the fact that matter and antimatter particles, on one side, and particles mediating the interactions, on the other, are described through the same objects, quantum fields, represents a first unification among all elementary particles, since their differentiation into matter–antimatter and interactions appears to be more secondary. This unification of particles into quantum fields is, in our opinion, the third most important unification in the realm of physical phenomena. The first was due to Isaac Newton, who unified the motions of heavenly and earthly bodies in free fall through his Law of Universal Gravitation: the force that makes an apple fall from a tree is the same force that makes the Moon revolve around the Earth and the planets revolve around the Sun. The second great unification was that of electricity, magnetism, and light (an electromagnetic wave), to which several scientists contributed. But it was mainly due to James Clerk Maxwell,[8] who published the complete set of 20 differential equations describing electromagnetism in his textbook *A Treatise on Electricity and Magnetism*, in 1873, six years before he died at the age of 48. Shortly afterwards, in 1884, Oliver Heaviside was able to reformulate the theory and reduce these equations down to four, and since them they are known as Maxwell's equations.

A crucial ingredient by which the Standard Model was obtained from Quantum Field Theory were the *symmetries* of the equations that describe the elementary particles. To give an example of what these symmetries mean, let us imagine that we write one equation that depends on the distance between two objects placed on a table. The equation will be the same if we bring that table to another room without moving the objects lying on its surface. In this case one says that the translation of the table is a symmetry of the equation because this one remains invariant when realizing that translation. Some of the symmetries of the Standard Model, which are called C, P, T, CP, and CPT, will be discussed in detail in Chap. 7 since they are relevant for the issue of the matter–antimatter asymmetry in the Universe. Here, we will just point out some interesting aspects about the most relevant symmetries of the Standard Model, known as *gauge symmetries*. They are related to the interactions between the elementary particles and they are only exact if the carriers of the interactions are massless. Otherwise, if the carriers of the interactions have a mass, one says that these symmetries are inexact or broken.

In the actual Universe, there is a gauge symmetry that corresponds to the strong interactions and another one that is related to electromagnetism. However, at the beginning of the Universe there was another symmetry that corresponded to the so-called electroweak interactions, resulting from the unification of the weak interactions with the electromagnetic ones. But this symmetry broke down in the first tenths of billionths of a second (10^{-10} s) after the Big Bang, that took place about 13,800 million years ago. This breakdown originated in the quantum field of the Higgs boson; to be precise,

---

[8]On the occasion of the centenary of Maxwell's birthday, Einstein described his work as the "most profound and the most fruitful that Physics has experienced since the time of Newton". When Einstein visited the University of Cambridge, in 1922, his host suggested Einstein had done great work because he stood on Newton's shoulders; Einstein replied: "No I don't. I stand on the shoulders of Maxwell".

it was due to the so-called *Higgs mechanism*, which generated masses for the W^+, W^- and Z^0 bosons. Moreover, the Higgs quantum field also provided masses for most elementary fermions: all six types of quarks, electrons, muons and tauons, providing identical masses for their antiparticles. We will return to this topic in Sect. 2.5.4 where we will take a closer look at the Higgs boson and the Higgs mechanism.

Gauge Groups of the Standard Model

The gauge symmetries of the Standard Model, associated to the interactions between elementary particles, follow the same patterns as some symmetry groups known in a mathematical discipline called Group Theory. As a consequence, in Particle Physics the terminology of that discipline is widely used and one says that the gauge symmetry of the Standard Model is given by the product of the groups:

$$SU(3) \times SU(2) \times U(1).$$

The gauge symmetry corresponding to the group SU(3) is related to the gluons, which mediate the strong interactions, and it is an exact symmetry because gluons are massless. The gauge symmetry corresponding to SU(2) $\times$ U(1) is related to the photons and the W^+, W^- and Z^0 bosons, all of them being the carriers of the electroweak interactions that unify the electromagnetic and weak interactions under the same framework. However, this symmetry was exact only at the starting of the Universe. Then, in less than a billionth of a second (10^{-9} s) after the Big Bang, it became inexact, i.e. broken, because the W^+, W^- and Z^0 bosons acquired masses, leaving a residual $U(1)_{em}$ symmetry corresponding to the electromagnetic interactions, which is exact because the photons are massless.

To conclude this brief overview on the genesis of the Standard Model, we would like to emphasize that the importance of the discovery of the Higgs boson at CERN, in 2012, was due to the fact that it confirmed the existence of the corresponding quantum field that provided the masses to most elementary particles. This discovery was the culmination of a whole epoch in Particle Physics, as the validity of the Standard Model was fully proven. However, this does not mean that this model is completely satisfactory. From the theoretical point of view, it contains too many free parameters, like the masses of the particles and the strength of the interactions, which cannot be deduced from first principles, so that their values are only known experimentally. Furthermore, the Standard Model does not account for all known phenomena related to Particle Physics, such as the existence of the enigmatic dark matter, and, in fact, it does not even describe the gravitational forces, as will be discussed in Sect. 2.6, devoted to several topics grouped under the name *Beyond the Standard Model*.

2.5.2 Elementary Fermions

Now we will review the main characteristics of the elementary particles in the Standard Model (Fig. 2.10), starting with the fermions. It happens that all known elementary fermions have spin 1/2 and are the particles which are considered to be matter (or antimatter). A curious and mysterious aspect of these elementary fermions is that they can be classified into three different families with identical properties except for their masses; to be precise, three families for the matter particles and another three "reversed" families for their antimatter counterparts. Each family of matter fermions consists of two quarks and two leptons, the latter being particles without strong or color charge, as explained above. One of the leptons has negative electric charge and the other has none and is called neutrino as it is electrically neutral. In what follows, first we will provide the electric charges and masses of the members of

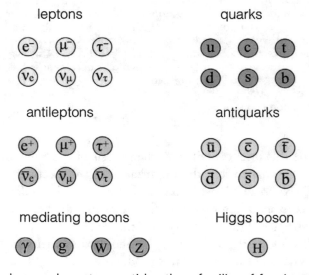

Fig. 2.10 The known elementary particles: three families of fermions, composed of two quarks and two leptons each, with their antimatter counterparts, plus the bosons mediating the interactions and the Higgs boson

these three families and then we will have a closer look at neutrinos, because of their many interesting features.

First Family

The first family of fermions is the lightest and its components enter the composition of atoms, except for the neutrino. It consists of the quarks u and d (*up* and *down*), the electron e^- and the electron neutrino ν_e. Their electric charges, taking as unit the proton's electric charge, are:

$$q(u) = 2/3, \quad q(d) = -1/3, \quad q(e^-) = -1 \quad \text{and} \quad q(\nu_e) = 0 \quad (2.5)$$

Notice that the quark charges are fractions of the proton's charge. The masses of the quarks, relative to the electron mass, are given approximately by:

$$m_u = 4.6\,m_e, \quad m_d = 10\,m_e, \quad \text{where} \quad m_e = 511 \text{ keV}/c^2. \quad (2.6)$$

Second Family

The second family of fermions is more massive than the previous one. It consists of the quarks c and s (*charm* and *strange*), with the same properties as the quarks u and d, respectively, plus the muon μ^- and the muon neutrino ν_μ. The masses of the quarks and the muon are approximately:

$$m_c = 2550\,m_e, \quad m_s = 190\,m_e \quad \text{and} \quad m_\mu = 207\,m_e \quad (2.7)$$

We see that the muon, otherwise identical to the electron, is 207 times more massive than this one. Its mean lifetime is about two millionths of a second: $\tau_\mu = 2.2 \times 10^{-6}$ s.

Third Family

The third family of fermions, the most massive, consists of the quarks t and b (*top* and *bottom*), the particle tau τ^- and the tau neutrino ν_τ. Their enormous masses reach the values:

$$m_t = 173 \text{ GeV}/c^2 = 346000\,m_e, \quad m_b = 8360\,m_e \text{ and } m_\tau = 3553\,m_e \quad (2.8)$$

The tau particle is 3553 times more massive than the electron, as we can see, and its mean lifetime is 10 million times shorter than that of the muon: $\tau_\tau = 2.9 \times 10^{-13}$ s. The top quark is the heaviest—the most massive— elementary particle that exists. With its 173 GeV/c^2, it is 346,000 times more massive than the electron and even heavier than the Higgs boson, whose mass is 125 GeV/c^2.

Antimatter Families

As was pointed out, to these three families of matter fermions one has to add the three families of antimatter fermions, with equal masses and mean lifetimes as their matter partners, but with opposite values for all charges. For instance, the first of these "antifamilies" is constituted by the antiquarks $\bar{u}$ and $\bar{d}$, the positron e^+ and the electron antineutrino $\bar{\nu}_e$. Their electric charges are therefore:

$$q(\bar{u}) = -2/3, \quad q(\bar{d}) = 1/3, \quad q(e^+) = 1 \quad \text{and} \quad q(\bar{\nu}_e) = 0 \quad (2.9)$$

Neutrinos and Antineutrinos

As for the masses of the neutrinos (and antineutrinos) of the three families, they are not known exactly, nor is the mechanism that generates them properly understood, although it is most likely that the field of the Higgs boson is involved in that mechanism. There are several methods used to obtain information about the neutrino masses, including astrophysical and cosmological observations that put very restricted upper bounds on the total sum of the three masses. Also, there are a number of laboratory experiments under way to directly determine the neutrino masses. The larger upper bounds found so far correspond to the "Mainz Neutrino Mass Experiment", from 2016, carried out in the University of Mainz (Germany), that results in $m_\nu <$ 2.2 eV/c^2 for the electron neutrino.[9] This bound is currently being tested by the KATRIN Experiment (Fig. 2.11), acronym for *Karlsruhe Tritium Neutrino Experiment*, conducted in the Karlsruhe Institute of Technology (KIT) in Germany, where the decay of tritium—the hydrogen isotope ^{3}H—is analyzed in the search for the electron antineutrino mass. In any case, taking into account other measurements, the three types of neutrinos are known to be more than 200,000 times lighter than the electron, the lightest fermion after the neutrinos, whose mass is 511 keV/c^2, as shown above.

An important observation is that the three neutrino species transform into each other during their propagation in free flight, that is, without colliding with any particles. To be precise, the three neutrino species, as well as the three antineutrino species, transform into one another but without mixing neutrinos with antineutrinos. This phenomenon, discovered in 1998 in the

---

[9]As a matter of fact, this upper bound was found for the electron antineutrino, which is the one involved in the experiments. Analogously, what the researchers of the KATRIN experiments can test is the mass of the electron antineutrino. Now, since the masses of neutrinos and their antiparticles must be identical, it is more convenient, for clarity, to refer to neutrinos rather than to antineutrinos.

Fig. 2.11 Spectrometer of the KATRIN Experiment in Leopoldshafen (Germany) in its way to the Karlsruhe Institute of Technology (Germany). It was constructed in Deggendorf in 2006 to be delivered only 400 km away. However, due to its almost 10 m width and its 200 tons weight, it could not be transported along the roads connecting the two towns and had to travel some 9,000 km to reach its destiny. This experiment is searching for the mass of the electron antineutrino analyzing the decay of tritium. *Credit* Courtesy of the Karlsruhe Institute of Technology (KIT)

Super-Kamiokande Neutrino Observatory in Japan (Fig. 7.8 in Chap. 7), is referred to as *neutrino oscillations* and because of them it is known that these particles have a mass. Indeed, in the Standard Model before 1998 it was assumed that neutrinos had no mass. These oscillations occur because the three neutrino species observed in the detectors, which interact with other particles through the weak interactions, are each a superposition of three basic neutrino states, called mass-eigenstates, which do not interact independently but propagate at different speeds since they have different masses. As a result, the superposition of the three basic neutrinos varies during the flight, giving rise, successively, to the three configurations that correspond to the three species of interacting neutrinos (called *flavor* neutrinos). This is a quantum mechanical effect and, as such, very difficult to understand with our intuition.

Neutrinos have a well-deserved reputation as ghostly particles, for they barely interact with ordinary matter, only through the weak and gravitational interactions.[10] This means that neutrinos (and antineutrinos) are practically invisible to the matter surrounding them and, conversely, the matter that surrounds neutrinos is practically transparent to them, so they can travel enormous distances through dense materials without being disturbed. Due to this property, it is not necessary to build tunnels in a neutrino factory, for example at CERN, to send neutrinos or antineutrinos underground to another laboratory. As a matter of fact, huge amounts of neutrinos are continuously crossing the Earth (and us) coming from diverse sources, mainly from nuclear reactions inside the Sun. These reactions produce electron neutrinos (but no antineutrinos), which are immediately released at almost the speed limit c. To be precise, the flux of *solar electron neutrinos* that hits the Earth each cm^2 per second is about 70,000 million. Moreover, the exposure to these neutrinos is the same regardless of whether it is day or night, as they pass through the Earth without hardly noticing its existence.

To the solar neutrinos one has to add the neutrinos and antineutrinos created by the natural radioactivity of the matter around us plus those originating in nuclear reactors, as well as those produced by cosmic rays when they collide with atoms in our atmosphere, known as *atmospheric neutrinos*. Furthermore, we also receive neutrinos and antineutrinos coming from beyond the Solar System, from stars and other astrophysical processes, especially the violent ones, like supernovae explosions. In addition, there is a cosmic *background radiation of neutrinos and antineutrinos* relic of the Big Bang, which escaped one to two seconds after it. But they have very low energies that make them very difficult to be detected, so they have not yet been found. Finally, it should be pointed out that *our bodies also emit neutrinos and antineutrinos*, due to the radioactive isotope of potassium ^{40}K, which forms part of our bones. This gives rise to some 4000 disintegrations per second, with one electron antineutrino being emitted in 89% of them and one electron neutrino in the remaining 11%.

There are other interesting issues related to neutrinos. In relation to anti-matter, they would be crucial to distinguish a hydrogen burning star, like our Sun, which emits neutrinos, from its antimatter counterpart—an antihy-drogen burning antistar—if it were to exist, which would emit antineutrinos, as is shown in detail in Chap.7, Sect. 7.1.1 (see Figs. 7.1 and 7.2). It is also curious that neutrinos move faster than photons, that is, faster than light, in practically all material media. This is due to the fact that light moves at the speed limit c only through the vacuum whereas neutrinos move very close

---

[10]However, at ultra high energies where the conditions of the first instant of the Universe are reproduced, the strength of the weak interactions is the same as that of the electromagnetic interactions, because they are essentially one and the same electroweak interaction, as explained above.

to the speed limit through all material media. Finally, we must point out that neutrinos might be their own antiparticles, as explained in Chap. 7, in Sect. 7.5 devoted to Ettore Majorana and his fermions. This possibility has been tested experimentally for several decades, although without any success so far, through various experiments aimed to detect the so-called *neutrinoless double beta decay*, which would unambiguously indicate that neutrinos are their own antiparticles (Majorana fermions). If this turned out to be true, some important long-standing problems of Particle Physics and Cosmology could be solved.

2.5.3 Elementary Spin 1 Bosons

Now let us examine the elementary bosons of the Standard Model and their principal characteristics. They are very different from the elementary fermions and are not distributed into families. To begin with, the spin 1 elementary bosons are the carriers that mediate the electromagnetic, strong, and weak interactions. Therefore, photons, gluons and the W^+, W^- and Z^0 bosons they all have spin 1.

Photons

Photons—the mediators of the electromagnetic interactions—represented by the letter γ, are very simple particles because they have no charges or mass, and are antiparticles of themselves. As they have no mass they move at the maximum possible speed c when they propagate through the vacuum, but not when they propagate through material media, as explained in Chap. 1. Another, and very important, consequence of photons having no mass is that electromagnetic forces are long range, as also happens with gravitation, and in both cases their strengths decrease with the square of the distance between the charges (electric charges or mass-energy "gravitational charges", respectively).

Although they have no mass, photons have energy which is proportional to the frequency of the associated electromagnetic wave and inversely proportional to its wavelength λ. Therefore, photons corresponding to different regions of the *electromagnetic spectrum* have different energies (see Fig. 2.12). The most energetic photons are the γ (gamma) rays[11] with the shortest wavelengths, $\lambda \leq 10^{-11}$ m, while the less energetic ones correspond to radio frequencies,[12] like radio and TV waves, with values of λ ranging from 1 m to

---

[11]The name gamma rays, or γ rays, was given by Rutherford to the photon radiation discovered at the end of nineteenth century in natural radioactivity, although at that time photons were not known to exist. Some decades later it was decided to represent photons, in general, with the same letter.

[12]Some experts classify microwaves as a subset of the radio waves, while others classify microwaves and radio waves as separate regions of the electromagnetic spectrum.

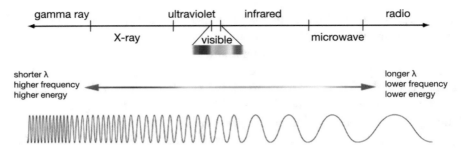

Fig. 2.12 Comparison of wavelength λ, frequency and energy for the electromagnetic spectrum. *Credit* NASA's Imagine the Universe

thousands of kilometers, being able to exceed 100,000 km in the case of ELF (*Extremely Low Frequency*) waves. Visible light is located in a middle region of the spectrum, with λ in a range of slightly less than one micrometer (1 μm = 10^{-6} m), also known as micron.

Curiously, photons and neutrinos (and antineutrinos) bear many resemblances in spite of the fact that photons are bosons and neutrinos are fermions. To start, photons are the most abundant known particles in the Universe along with neutrinos and antineutrinos. Indeed, it is estimated that for each proton there are about 1.6×10^9 photons and a slightly lower number of neutrinos and antineutrinos. However, there could also exist particles that we do not yet know, such as dark matter particles, more abundant than photons. Another similitude is that large quantities of photons and neutrinos are created in the nuclear reactions that take place inside the stars. The electron neutrinos, to be precise, leave the star immediately at almost the speed limit c, as explained above. By contrast, the emitted photons get stuck inside the star and it will take hundreds of thousands of years for them to reach the surface and come out as light; mainly visible, infrared, and ultraviolet light. Similarly, a multitude of astrophysical phenomena release huge quantities of photons, neutrinos, and antineutrinos. Furthermore, there are two relic background radiations from the Big Bang, one of photons and the other of neutrinos and antineutrinos, and they permeate the entire space in all directions evenly. The neutrino radiation dates from one to two seconds after the Big Bang, while the photon radiation was released 380,000 years later, once the simplest atoms had formed.

We see therefore that neutrinos escape much faster than photons from the interior of stars, and they also escaped much sooner from the influence of the other particles in the early Universe. This happens because photons interact substantially with the charged particles that make up matter and, consequently, they often have difficulty propagating through material media.

This situation reminds of a lady who wants to walk through a large salon with a multitude of guests chatting animatedly in small groups. If the lady can barely see them, nor can the others see her, she will keep going forward without pausing; but if she knows most of the guests and they know her and enjoy a good relationship, then she will slow down to chat with everyone. As a result, it will take her a long time to cross the room.

Gluons

Gluons—the mediators of the strong interactions—are far more complex than photons and, like them, have no mass. To begin with, there are eight different types of gluons versus a single type of photon. In addition, gluons also carry strong or color charge, consisting of a color-anticolor pair; therefore, they interact strongly not only with quarks and antiquarks but also with other gluons and with themselves. This is as if while playing ball (the players are two quarks and the ball a gluon that they interchange) other balls would suddenly emerge from that ball, which could even form compounds of two or more balls called *glueballs*. As a matter of fact, the Standard Model predicts the existence of glueballs accessible to current accelerators, but they have not been identified yet in a conclusive way. This is due to the fact that glueballs look very similar to mesons and it is very difficult to tell them apart.

As if this were not enough complication, gluons also produce the confinement of strongly charged particles, i.e. the confinement of quarks, antiquarks, and themselves. This is technically called *color confinement* and implies that these particles cannot be separated individually; that is, no single quarks or antiquarks can be obtained but they always come together with other quarks and/or antiquarks in order to neutralize their colors so that the resulting composite particle is neutral, as discussed in Sect. 2.4.3. This is so because the strength of the strong force increases with the distance between the particles, just as the force between two portions of an elastic band increases as we pull them apart. Consequently, if the quarks and/or antiquarks are very close to each other they hardly notice any "strong" attraction, although they certainly still notice the electrostatic attraction or repulsion due to their electric charges, and they can even feel the weak interactions. But if they try to move away from each other individually then they will notice the force that attracts them towards other quarks and antiquarks. The confinement of gluons, on the other hand, has the effect that they cannot propagate freely through space, even though they have no mass.

The W^+, W^- and Z^0 Bosons

The W^+, W^- and Z^0 bosons—the mediators of the weak interactions—are very different from photons and gluons. The first two have opposite electrical charges, being antiparticles of each other, while the Z^0 boson is neutral and antiparticle of itself. They are quite massive, and their mean lifetimes are so short (of the order of 10^{-25} s) that they only travel very small distances, of about 10^{-18} m, before they decay. Their masses are: $m_W = 80.4$ GeV and $m_Z = 91.2$ GeV.

As was explained in Sect. 2.4.4, weak interactions are responsible for most decays of particles, and in particular for the neutron β-decay, both inside and outside the atomic nuclei. This decay, which results in the β radiation of electrons, occurs according to the process:

$$n \to p + e^- + \overline{\nu}_e, \tag{2.10}$$

in which the neutron "becomes" a proton, emitting one electron and one electron antineutrino. When the neutrons are free, i.e. outside the atomic nuclei, they become more unstable than inside them, and decay with a half-life of about 10 min. This means that if we have any number of free neutrons, after 10 min half of them will have disintegrated.

The β-decay of the neutron as expressed in (2.10) does not show the role played by the mediators of the weak interactions. What actually happens, looking at shorter distances, is that a quark d of the neutron emits a W^- boson, so that the quark d "becomes" a quark u, resulting in the neutron being transformed into a proton. This W^- boson decays immediately, in turn, giving as by-products the electron e^- and the electron antineutrino $\overline{\nu}_e$. So, as shown in the Feynman diagram of Fig. 2.13, the neutron β-decay consists in reality of two phases:

$$d \to u + W^- \quad \text{and} \quad W^- \to e^- + \overline{\nu}_e \tag{2.11}$$

giving $d \to u + e^- + \overline{\nu}_e$, which finally translates into the observed process (2.10). In this type of diagrams, invented by Richard Feynman (Fig. 2.8), time goes from left to right. Curiously, the lines corresponding

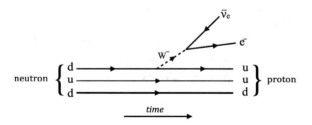

Fig. 2.13 Feynman diagram corresponding to the β-decay of the neutron (u,d,d), resulting in a proton (u,u,d) plus one electron e^- and one electron antineutrino $\overline{\nu}_e$

to antifermions are indistinguishable from those corresponding to fermions but are always drawn with the arrows pointing in the opposite direction of their motion, as if the antimatter particles were the ordinary matter particles traveling backwards in time.[13]

A very peculiar and curious property of weak interactions is the fact that by emitting, or receiving, W^+ and W^- bosons, the particles change their identity "turning" into other different particles. For example, as we have just seen, in the neutron β-decay a quark d emits a boson W^- thus becoming a quark u. This phenomenon does not occur with any other interactions, nor even with the exchange of the Z^0 boson. Let us see this in more detail.

When a particle with electric charge emits or receives a photon γ the particle feels the corresponding attraction or repulsion, but it is still the same particle, its identity does not vary. The same can be said of a particle with strong charge or color, although with the subtlety that in this case the color of the particle does change when it emits or receives a gluon g. This is like if our particles were swimmers playing ball on the beach. If the ball is a photon, nothing out of the ordinary happens playing with it, but if the ball is a gluon the color of the swimsuits changes every time the players throw the ball or pick it up. The most extraordinary and weird happens, however, when the ball is one of the W^+ and W^- bosons, since in that case the swimmers become other swimmers that were not even playing ball! This magic is explained by the Quantum Field Theory that describes elementary particles because in this theoretical framework the creation and annihilation of particles occurs in a systematic way.

The weak interactions mediated by the W^+ and W^- bosons are also responsible for the processes of nuclear fusion that take place inside hydrogen burning stars, such as our Sun, so we have to be very grateful to these interactions for the light and heat that they provide, making the support of life possible, among other virtues.

The sequence of nuclear reactions in the Sun that culminates in the fusion of hydrogen producing helium is called the *proton-proton chain reaction*. It starts with the union of two protons resulting in a deuterium nucleus (one proton and one neutron) plus one positron e^+ and one electron neutrino ν_e. That is:

$$p + p \rightarrow p + n + e^+ + \nu_e \tag{2.12}$$

---

[13] Feynman, in his 1949 article "The Theory of Positrons", cited Ernst Stückelberg for the idea of picturing the positrons in spacetime as electrons traveling backwards in time.

Thus, one of the protons is transformed into a neutron through a process, so to speak, "inverse" of the neutron β-decay and, like this one, it takes place in two phases. First a quark u emits a W^+ boson becoming a quark d: $u \rightarrow d + W^+$, whereby the proton is transformed into a neutron, and then the W^+ boson decays giving the positron e^+ and the electron neutrino ν_e. As a consequence, the Sun and all hydrogen burning stars produce enormous amounts of antimatter in the form of positrons as well as enormous amounts of electron neutrinos. The positrons immediately annihilate with the electrons around, producing photons of light and heat, whereas the neutrinos leave the star very quickly at almost the maximum speed c, as we know. We will return to this in Chap. 7.

2.5.4 The Higgs Boson and the Higgs Mechanism

The Higgs boson, with spin 0, has no charges, so it is its own antiparticle. It is very heavy, 133 times more massive than the proton, its mass being: $m_H = 125$ GeV. Its mean lifetime is given by: $\tau_H = 1.56 \times 10^{-22}$ s. It is therefore an incredibly fleeting particle, although still 1,000 times more long-lived than the carriers of the weak interactions. This extremely short lifetime made many curious people wonder, following the discovery of the Higgs boson, how could this particle be so important for Particle Physics and the Universe if it could hardly be said to exist. And these people were not without reason since the Higgs boson discovered at CERN has a very limited relevance for Particle Physics. The importance of that discovery, as was pointed out before, lies in the fact that it confirmed the existence of the Higgs quantum field, which is responsible for providing the mass to most elementary particles. Let us see this in more detail.

As was explained in Sect. 2.5.1, the gauge symmetry associated to the electroweak interactions that existed at the beginning of the Universe, unifying the electromagnetic and weak interactions, was broken in the first tenths of billionths of a second after the Big Bang. As a consequence, electromagnetic and weak interactions were separated and became independent from each other. This occurred because the Higgs quantum field acquired a non-zero value in its state of minimum energy—the vacuum—unlike the other quantum fields (Fig. 2.14), which remained hidden in the vacuum with a zero value and only manifest themselves through quantum fluctuations.[14]

---

[14]The potential energy shown on the left of Fig. 2.14 is a simplification of the true potential energy of the Higgs field, which is a four-dimensional "surface" instead of a one-dimensional line. This is so because the Higgs field had four components at the beginning of the Universe; that is, there were four Higgs fields, but only one survived the electroweak symmetry breaking. It should also be noted

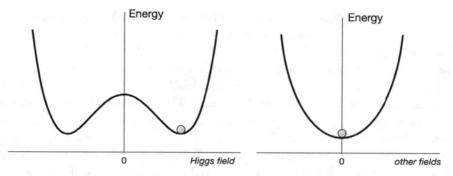

Fig. 2.14 The potential energy of the Higgs field (left) and of the other quantum fields (right) as a function of their values. While the other fields have a zero value in the state of minimum energy, indicated by the little ball, the Higgs field has a non-zero value and fills the Universe with that value. This generates the masses of most elementary particles, which in turn allows the existence of atoms

This phenomenon gave rise to the *Higgs mechanism,* which generated masses for the W^+, W^- and Z^0 bosons, causing the breakdown of the electroweak symmetry as a result. The reason for this is that gauge symmetries are only exact symmetries for interactions mediated by massless bosons, as said before. Furthermore, the Higgs field also generated masses for all the quarks, for the charged leptons e^-, μ^- and τ^- and for the Higgs boson itself, as well as for all their antiparticles, but this was not accomplished through the Higgs mechanism, but via the so-called *Yukawa couplings.*

Although there were precursors of the concept of the Higgs mechanism in Condensed Matter Physics, which is non-relativistic, the relativistic version for High Energy Physics was developed and put forward in 1964 by three independent groups: first by Robert Brout and François Englert, from the Université Libre de Bruxelles (Belgium), two months later by Peter Higgs, at the University of Edinburgh (UK), and a few months later by Gerald Guralnik, Carl Hagen and Tom Kibble, from Harvard University (USA), Rochester University (USA) and the Imperial College at London (UK), respectively. Anyway, it is noteworthy to point out that the prediction of the existence of a new particle, which years later was dubbed "the Higgs boson", was only in an article written by Peter Higgs, but not in any other article. Although the existence of that particle could be deduced from the contents of some of the other articles, the real fact is that this possibility was not mentioned in those articles.

---

that in Particle Physics, as well as in Cosmology, various hypothetical quantum fields have been proposed, for different reasons, which would have non-zero values in the state of minimum energy, like the Higgs field.

In 1967 Abdus Salam, at Punjab University (Pakistan), and Steven Weinberg, then a visiting professor at MIT (USA), implemented the Higgs mechanism to the Standard Model, applying it to the breaking of the electroweak gauge symmetry and the generation of the masses for the W^+, W^- and Z^0 bosons. They received the 1979 Nobel Prize in Physics, along with Sheldon Glashow, from Harvard University (USA), who had introduced the Z^0 boson and had proposed the correct gauge symmetry for the electroweak interactions. Ironically, even though part of that award was given for implementing the Higgs mechanism, it still took 34 years until, in 2013, François Englert and Peter Higgs (Fig. 2.15) received the Nobel Prize in Physics, after the discovery of the Higgs boson in 2012 at CERN (Brout had already died).

The Higgs mechanism and the generation of the masses via the Yukawa couplings are relativistic quantum mechanical effects that originate from the non-zero value of the Higgs field at its minimum energy, that is, in the vacuum. Both these effects are deduced from the equations of the Standard Model, but they are far from intuitive, again, as they belong to the quantum mechanical realm. In this respect, one of the most popular analogies, put forward to "explain" intuitively the generation of masses through the Higgs mechanism, is that the Higgs field creates a kind of viscosity in the Universe,

Fig. 2.15   Peter Higgs around 2012 with some LEGO models of the ATLAS detector of the Large Hadron Collider (LHC) at CERN. ATLAS and CMS were the two detectors that registered the creation of the Higgs boson at the LHC. *Credit* Courtesy of CERN

so that particles feel a certain resistance to move through it, like an inertia which would translate into their masses. The trouble with this analogy is that "true" masses, true inertia, only present resistance to accelerations, not to uniform velocities, and therefore the masses of the particles do not manifest themselves when the particles move through space at a constant speed; only when they are accelerated. So, no viscosity can explain the true facts about the mass as the cause of inertia.

One obvious question now is why the Higgs field behaved so strangely, why it developed a non-zero value in the vacuum—the state of minimum energy—unlike the other quantum fields. The answer is that before the breakdown of the electroweak symmetry the value of the Higgs field in the vacuum was zero, as for the other quantum fields, and no elementary particle had a mass. But the temperature of the Universe dropped as it expanded, and this caused a phase change for the Higgs field, like when water cools down and ice crystals start to form. Then bubbles of the new phase began to emerge, in which the value of the Higgs field was non-zero, and the particles inside the bubbles acquired a mass, as shown in Fig. 2.16. Very quickly more and more particles moved inside the bubbles, which were growing relentlessly, until finally the bubbles filled the entire space: the phase transition was completed.

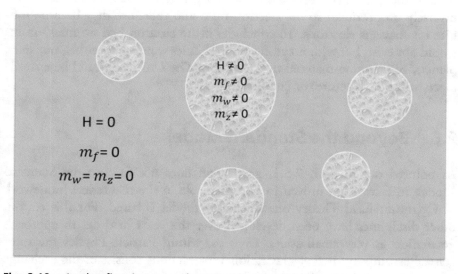

Fig. 2.16 In the first instants after the Big Bang there was a phase change for the Higgs field due to the drop in temperature in the Universe as it expanded. Then bubbles of the new phase began to emerge, in which the value of the Higgs field was non-zero, $H \neq 0$, and the particles inside the bubbles acquired a mass: m_f represents the masses of the fermions while m_w and m_z are the masses of the W and Z bosons, carriers of the weak interactions

An important remark about the masses generated by the Higgs field is that they contribute to the mass of atoms only slightly more than 1%. The remaining 99% comes from the energy of the strong interactions within the atomic nuclei, due to the mass-energy equivalence $E = m c^2$, although the mechanisms behind this mass generation are not well understood yet and are object of intense research. Indeed, knowing that the masses of the quarks u and d are: $m_u = 2.3$ MeV/c^2 and $m_d = 4.8$ MeV/c^2; and taking into account that the proton mass is $m_p = 938.3$ MeV/c^2, we obtain that the sum of the masses of the three quarks (two u and one d) is equal to 9.4 MeV/c^2, only 1% of the mass of the proton. In the case of the neutron the percentage is a bit higher as the sum of the masses of the three quarks (two d and one u) gives 11.9 MeV/c^2, while the neutron mass is $m_n = 939.6$ MeV/c^2.

Nevertheless, although the action of the Higgs field produces such a small percentage of the atom mass, and therefore of the mass of all the matter around us, that contribution is crucial because if the electron had no mass, atoms would not exist. This is so because massless electrons would move at the speed limit c making it very hard for protons to capture them. Furthermore, if some electrons were caught for a brief moment, they would not bind to the protons to form atoms because the radii of the atomic shells are inversely proportional to the electron mass and therefore they become infinite for massless electrons. To conclude, if the electron had no mass, atoms would not exist, but without atoms no molecules would exist either, nor stars, planets, or anything material that we know. The Universe would be a sterile place, without structures or complexity.

2.6   Beyond the Standard Model

As pointed out in Sect. 2.5.1, gravitation does not fit into the Standard Model; in fact, it has proved to be intractable in the theoretical framework of Quantum Field Theory on which this model is based. For this reason, other disciplines have been developed over the years in order to approach gravitation at very small scales. Even so, within Particle Physics the most general assumption is that gravitational interactions consist of the exchange of a hypothetical spin 2 boson called graviton, in a similar manner to the other interactions. In fact, one of the research areas in Particle Physics, called String Theory, is currently the only framework that accounts for gravitation and gravitons together with the other elementary particles and forces. This theory assumes that all known particles are one-dimensional entities similar to tiny strings, and unifies the Standard Model with gravitation, although not without controversy. Yet there are other theoretical frameworks, including several very groundbreaking ones, that consider gravitation not as a fundamental force, but as an emergent phenomenon from collective properties of matter, such as entropy, temperature, or even quantum entanglement.[15]

As regards the search for elementary particles, it has not ended with the finding of the Higgs boson at CERN. The search continues at present and will continue in the future with increasingly powerful accelerators. This is so because the Standard Model, despite its successes, has also many aspects that can be improved, as noted before. This has motivated the study and development of theories and models that go beyond the Standard Model and predict the existence of new elementary particles, as is the case in String Theory. As a result, various types of particles are actively searched for, which, for one reason or another, are thought to exist. For example, more Higgs bosons are searched, besides the single one of the Standard Model, because most theoretical models predict at least two "higgses". Likewise, the search for supersymmetric particles, Kaluza-Klein particles, axions, as well as dark matter particles, is also very intensive. Let us have a closer look at these hypothetical particles.

2.6.1 Supersymmetric Particles

The property called *supersymmetry*, that relates bosons with fermions, was initially discovered as a possible symmetry of Quantum Field Theory by several independent groups in the early 1970s. These groups were exploring the symmetries allowed in that formalism, regardless of any phenomenological applications to the real world. Then, in 1974, supersymmetry became relevant for Particle Physics after an article appeared by Julius Wess (Fig. 2.17), from Karlsruhe University (Germany), and Bruno Zumino, at CERN. The application of supersymmetry to the Standard Model started only one year later by Pierre Fayet, at L'Ecole Normal Superieure, in Paris (France), and many articles followed, subsequently, by the same and other authors, due to the remarkable potential of supersymmetry for solving long-standing puzzles in the Standard Model. Fayet also introduced the idea of supersymmetric partners (see below). Some years later this framework

---

[15]The entropy of a physical system can be regarded as the disorder of its components; the temperature measures its thermal energy; that is, the average kinetic energy of the thermal fluctuations or vibrations of its components; and quantum entanglement is a distinctive quantum property for which two or more components may share the same quantum state even if they get spatially separated by large distances.

Fig. 2.17   Julius Wess (1934–2007) in the Mathematisches Forschungsinstitut Ober-wolfach (MFO, Germany) in 2006. Together with Bruno Zumino he wrote down in 1974 the first supersymmetric quantum field theory, called the Wess-Zumino model. With Jonathan Bagger he published the book *Supersymmetry and Supergravity*, in 1983. The Julius Wess Award, dedicated to his memory, is granted by the Karlsruhe KCETA (Germany) to particle physicists for outstanding scientific achievements. *Credit* Renate Schmid, Courtesy of MFO

received an extra impetus from String Theory, as most of its "realistic" models were supersymmetric.

To be short, if in the realm of High Energy Physics supersymmetry existed, then the number of elementary particles would double because for every known ordinary particle there would be a partner with the bosonic-fermionic character inverted. These partners of the ordinary particles are dubbed *super-symmetric*, or *SUSY*, particles. To be more precise, for each elementary spin 1/2 fermion there would exist a spin 0 boson, with the same name but preceded by an "s"; and for each elementary boson there would exist a spin 1/2 fermion, with the same name but ending in "ino". Accordingly, if super-symmetry were a reality, then s-electrons, s-positrons, s-neutrinos s-muons and s-quarks would exist together with fotinos, gluinos and higgsinos, among other SUSY particles, as one can see in Fig. 2.18.

Of special interest would be the *lightest supersymmetric particle* (LSP) because it could be stable and have the right properties to qualify for "the dark matter particle" which would account for most dark matter in the Universe (see below). Now, although the LSP might be the lightest s-neutrino or the gravitino—the SUSY partner of the graviton—the most likely possi-bility, according to the experts, is that it would be one of the so-called

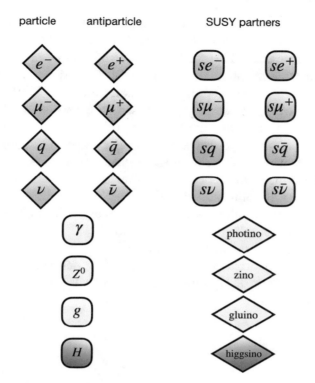

Fig. 2.18 Most Standard Model particles and antiparticles with their SUSY partners with the bosonic and fermionic character inverted (q stands for quarks and sq for s-quarks). Fermions are depicted as rhombuses and bosons as rounded squares

neutralinos, spin 1/2 fermions with no electric charge which are quantum mixtures of the higgsino with the SUSY partners of the electroweak gauge bosons. Neutralinos would be their own antiparticles, so they would annihilate with themselves, and their masses would be typically between 100 and 1000 GeV, therefore much larger than those of most Standard Model particles.

2.6.2 Kaluza-Klein Particles

Kaluza-Klein particles, named after the mathematician Theodor Kaluza and the physicist Oskar Klein, would only appear if there existed other space dimensions besides the three that we perceive. In fact, these particles would consist of the ordinary particles but moving also through, at least, one extra dimension in addition to the three ordinary ones. This would have the effect of provide these particles with a larger mass once they were detected. There-fore, the discovery of a Kaluza-Klein particle would confirm the existence of extra space dimensions. There are a number of reasons in Particle Physics for considering the possibility of other spatial dimensions; in fact, this idea has been around for over a century in Physics. Indeed, in 1919 Kaluza, from Königsberg University (Germany), proposed an extension of General Rela-tivity with one extra space dimension in order to unify gravitation with electromagnetism, an idea that eventually got Einstein's approval. And in 1926 Klein, from the University of Copenhagen (Denmark), proposed that Kaluza's extra dimension was curled up into itself and was microscopic; in other words, it was compactified with a very small radius (see Fig. 2.19). Einstein studied the resulting Kaluza-Klein theory with interest a few years later, especially during the period 1938–1943.

Some decades later, the original Kaluza-Klein theory was abandoned once it became clear that electromagnetism could be best described in the quantum relativistic framework of Quantum Field Theory, together with the newly

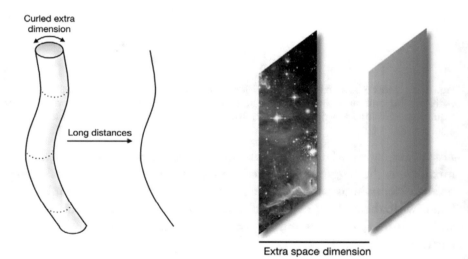

Fig. 2.19 On the left, there is an example of Kaluza-Klein compactification, where one dimension is much smaller than the others and curled up into itself. At long distances, a two-dimensional surface with one such small dimension looks one-dimensional. On the right, one can see the setup of the RS-I model, consisting of two three-dimensional branes, which are parallel along one extra space dimension which is compactified. One of the branes represents our Universe while the other is depleted of matter or antimatter but is a major source of gravitation. This model can explain the weakness of gravity in our Universe in terms of a "warped" geometry along the extra dimension, whose length is extremely short and does not extend beyond the two branes

discovered strong and weak interactions, but the same was not true for General Relativity, despite many efforts to integrate it as well. It should be noted, however, that many failed attempts were also made in order to describe the new interactions within the same Kaluza-Klein framework but with additional extra space dimensions. Even so, in more recent times the idea of extra space dimensions acquired renewed interest due to the advent of String Theory. In this new framework, which passed unnoticed in the 1970s, but was embraced by hundreds of physicists all over the world in the middle 1980s, elementary particles are described as tiny strings that live in spaces of more than three dimensions. Now, in String Theory there is a plethora of additional particles and several of them have been proposed as dark matter candidates: mainly SUSY partners, but also some particles of Kaluza-Klein type, as well as a variety of axion-like particles (see below).

Besides String Theory, many research groups have also postulated extra dimensions and the resulting Kaluza–Klein (KK) particles. For example, the Randall-Sundrum (RS) models, and their many variations, add typically one extra dimension along which three-dimensional "branes" (generalization of two-dimensional membranes) are located, parallel to each other. Some of these branes represent worlds or universes, like ours, but others do not, although they influence the brane universes, and their presence is needed to explain some physical properties of the latter. In Fig. 2.19 one can see the setup of the RS-I model, proposed in 1999 by Lisa Randall (Fig. 2.20), then from Princeton University (USA), and Raman Sundrum, from Boston University (USA). It consists of only two three-dimensional branes, along the extra space dimension. One of the branes represents our Universe while the other is depleted of matter or antimatter, but is a major source of gravitation and can explain its weakness in our Universe in terms of a "warped" geometry along the extra dimension. This means that moving through the extra dimension the mass and size of any particle change. The extra dimension, whose size is extremely small, is compactified and does not extend beyond the branes. This model predicts KK partners of the gravitons, whose signatures could be accessible to the LHC at CERN. In other similar models, some KK partners of the gauge bosons mediating electroweak interactions have been proposed as dark matter particles too.

2.6.3 Axions

Axions are spin zero particles predicted to exist more than 40 years ago in order to solve some inconsistencies between theory and observation in the realm of strong forces, known as the *strong CP problem*, where *CP* is the

Fig. 2.20 Lisa Randall is Professor of Science at Harvard University (USA), working in Particle Physics and Cosmology. In this image she is giving a TED talk (*Technology, Entertainment, Design*) on the topic of extra dimensions, in 2006. Apart from her scientific work on extra dimensional brane worlds, especially the Randall-Sundrum I and II models, she has also become widely known for her popular scientific books. *Credit* Courtesy of Lisa·Randall

operation that transforms particles and antiparticles into each other, and is an almost exact symmetry in the Standard Model.[16] This problem has to do with the non-observation of the violation of this symmetry in processes involving strong interactions, violation that is observed in some processes involving weak interactions, taking into account that the theory that describes the strong interactions permits such violation as well. Namely, this theory contains a parameter, called θ, that must be extremely small, at least ten orders of magnitude smaller than its natural value, in order for strong interactions to preserve the CP symmetry at the current level of accuracy.

The term axion was coined in 1978 by Frank Wilczek, from Princeton University, who predicted its existence along with Steven Weinberg, at Harvard University. This hypothetical particle, which is its own antiparticle, was a natural outcome of a new quantum field associated to the θ parameter, that had just been postulated by Roberto Peccei and Helen Quinn, both at Stanford University (USA), in order to solve the strong CP problem. The prediction of the axion has a rather intricate history and its predicted mass has changed substantially over time, as well as its nature, so instead of "the

---

[16]In Chap. 7 we will discuss the CP symmetry in detail, including its violation by the weak interactions.

axion" many axion-like particles are nowadays referred to under the generic name of axions.

In 1983 Pierre Sikivie (Fig. 2.21), at the University of Florida (USA), studied the theoretical coupling of the axion to the electromagnetic field and he realized that axions could be converted into photons by means of a strong inhomogeneous magnetic field. He also made some estimates about the flux of axions emitted by the Sun and about the axion abundances in the halo of our galaxy. This led to several experiments that are presently underway. For example, there is the *CERN Axion Solar Telescope* (CAST) experiment aimed at the Sun, as our star is supposed to be a good source of axions which the CAST experiment expects to turn into X-rays using a strong magnetic field of up to 9.5 Tesla.

Most recently, in June 2020, the XENON collaboration has reported the possible discovery of solar axions in the XENON1T experiment, at the Gran Sasso National Laboratory (Italy). But the results can also be explained by β decays of traces of tritium (^{3}H), which are contaminating the detector. In addition, it was pointed out that the data are inconsistent with bounds on axions from the cooling of white dwarf stars.

Fig. 2.21 Pierre Sikivie—dubbed "the Axionman" by some of his colleagues—is a Distinguished Professor of Physics at the University of Florida (USA). This photo was taken during an Axion Training workshop at CERN, in 2006. He explained the technique he invented in 1983 to detect axions, on which all magnetic axion telescopes, like CAST, are based. In 2019 he was awarded the APS Sakurai Prize, one of the Physics highest honors. *Credit* Courtesy of Pierre Sikivie and CERN

2.6.4   Dark Matter Particles

Dark matter is a proposal to solve several astrophysical conundrums, such as the rotation speed of spiral galaxies and ghost gravitational lenses. In brief, it happens that in the Universe there seems to be much more matter than observed, judging by the gravitational fields detected, and this invisible matter does not emit any electromagnetic radiation; at least, radiation that we are capable of detecting. Among many ideas that have been put forward in order to solve the dark matter mystery, there is the proposal that most dark matter is made of particles not discovered yet, that lie beyond the Standard Model. Accordingly, many candidates have been postulated to be the dark matter particles, including some of the hypothetical particles that we have just reviewed in the paragraphs above, such as SUSY neutralinos, KK particles and axions. It must be said that neutralinos and the lightest KK particles are examples of the so-called Weakly Interacting Massive Particles (WIMPs), which are sensitive to only the weak interactions—or even weaker interactions—and are searched for in many experiments aimed at the detection of dark matter. In fact, neutralinos with masses between 10 and 10,000 GeV are the favorite WIMP dark matter particles.

Axions are too light to be considered as WIMPs. Nevertheless, it seems that axions with masses above 10^{-11} times the electron mass could also fit into the description of dark matter particles. Thus, they could also be discovered in some experiments designed for the detection of dark matter particles, like the experiments carried out by the aforementioned XENON collaboration, which is searching for dark matter since 2006 using liquefied gas xenon.

In any case, dark matter, along with dark energy is the main topic of the next chapter, so we will continue giving more details there.

3

Antimatter Versus Dark Matter and Dark Energy

Antimatter, dark matter and dark energy constitute the exotic side of the Universe. Even though they are completely distinct "substances", antimatter is often confused with dark matter, which in addition is also confused with dark energy. The aim of this chapter is to clarify these misconceptions and, in so doing, to enlighten the readers on the important topics of dark matter and dark energy. First, we will briefly review the discovery of the expanding Universe and the genesis of the Big Bang theory. Then we will describe the key features of antimatter, dark matter and dark energy in a few sentences, including their contribution to the total matter of the Universe and to its energy density. The composition of the Universe today with all its components will follow. After that, we will have a closer look at dark matter and dark energy in two sections. For readers not acquainted with atomic spectroscopy and spectral lines there is an introduction to the subject in Appendix A.

3.1 The Expanding Universe

The Universe came into existence some 13,800 million years[1] ago out of a sudden expansion of the space—the *Big Bang*—starting from an initial state of extremely high density and temperature. The space expanded carrying all

---

[1]That is 13.8 billion years. In English, billion = thousand million = 10^9 while in many European languages billion = million million = 10^{12}. The age of the Universe is not known exactly since there are some discrepancies in the values of the observed cosmological parameters that determine it using different techniques. Its most updated estimate in the Standard Model of Cosmology is 13,787 ± 200 million years.

© Springer Nature Switzerland AG 2021
B. Gato-Rivera, *Antimatter,*
https://doi.org/10.1007/978-3-030-67791-6_3

its matter-energy content along, like an expanding fireball, and it continues to expand today. However, the matter-energy content from the early Universe was very different from that of the present Universe and was undergoing continuous transformations as the space expanded. For example, atoms only appeared about 380,000 years after the Big Bang, whereas the first stars and galaxies still needed some 400 million years to form.

One may wonder what was there before the Big Bang, of course. The honest answer to this question is that nobody knows although there is a whole variety of different proposals. These include a number of cyclic models in which the Universe undergoes successive cycles, each cycle starting with a Big Bang and ending with either a *Big Crunch* or a period of ultra-slow contraction, named *ekpyrotic phase*, among other possibilities. The idea of cyclic universes is not the most widely-held view among cosmologists, however. The majority of them believe in the hypothesis that the Universe underwent a phase of extremely rapid expansion at the very beginning and during an extremely short time, called *cosmological inflation*. If this proposal turns out to be true, then all information about anything previous to that phase was washed out, erased for all practical purposes. But it must be stressed that the inflationary hypothesis, that dates back to 1981 and has given rise to many different models, has not yet been proven.

The Big Bang theory, on the contrary, has been proven beyond any reasonable doubt by now, and by different independent methods. It was put forward for the first time in 1931 by the Catholic priest Georges Lemaître (Fig. 3.1), although in a rather rudimentary form called *hypothesis of the primeval atom*. A few years earlier, in 1927, Lemaître had proposed a model of an expanding Universe in order to explain the *redshift* of the light coming from most spiral nebulae, as spiral galaxies were named at that time.

The spectral shift of the light emitted by an astronomical body, usually denoted as z, depends on the speed of the body with respect to the Earth due to the *Doppler effect*,[2] as shown in Fig. 3.2. In particular, the redshift ($z > 0$) in the spectral lines of most galaxies means that the wavelength λ of the light emitted by them has increased because such galaxies are in recession, i.e. moving away from us; and the opposite is true for the light from the few galaxies coming towards us: λ has decreased showing a blueshift ($z < 0$). Now, as we will see in Sect. 3.5.1, in an expanding Universe the redshifts of

---

[2]The Doppler effect consists of the change of the wavelength λ, for any type of wave, produced by the relative motion between the body emitting the wave and the receiver. It is what produces an appreciable sound difference in the whistle of a train, depending on whether it approaches or moves away from us. The optical spectral shift z is the relative variation of the wavelength λ of the light emitted by the source with respect to the λ_0 received at our telescopes: $z = (\lambda_0 - \lambda)/\lambda$. Positive values, $z > 0$, correspond to redshifts and negative values, $z < 0$, to blueshifts.

Fig. 3.1 Georges Lemaître (1894–1966) around the mid-1930s. He was ordained a priest of the Catholic Church in 1923 and was Professor at the Catholic University of Louvain (Belgium). In spite of this, he never tried to mix science with religion, and he did not find any conflict between them. He was the first scientist to realize that the recession of most galaxies could be explained by an expanding universe, an idea that Einstein thoroughly rejected at first, even though his equations of General Relativity implied that the Universe was either expanding or contracting. *Credit* Courtesy of Wikimedia Commons

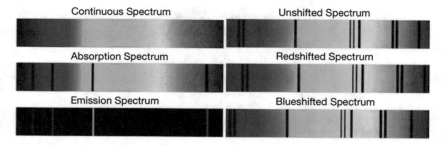

Fig. 3.2 On the left, a continuous light spectrum together with spectral lines caused by chemical elements absorbing or emitting photons at specific wavelengths. On the right, spectral lines shifted due to the Doppler effect. In the case of far away galaxies, the redshifts are mainly due to the expansion of the Universe. *Credit* Courtesy of Jhausauer and Wikimedia Commons (left) and NASA/JPL-Caltech (right)

far away galaxies are more appropriately understood as a consequence of the stretching of space behind the Universe's expansion.

The expanding Universe that Lemaître proposed in 1927 was a solution he had found to Einstein's equations of General Relativity while he was working on his thesis at MIT (Massachusetts Institute of Technology, USA). However, it was an eternal Universe, with no beginning or end. Hence, it still took him four years to propose an expanding Universe with a beginning and (possibly) an end, which was another solution to Einstein's equations. This new solution was a rediscovery, in fact, since it had been found earlier, in 1922, by the Russian mathematician Alexander Friedmann, who unfortunately died very young—at the age of 37—in 1925. As for Einstein, he totally rejected the idea of an expanding Universe until the evidence became overwhelming.

But let us go back to 1912, the year in which modern Cosmology was born and crucial astronomical observations were made, which eventually gave rise to the theory of the expanding Universe and the Big Bang (Fig. 3.3). First of all, Henrietta Leavitt, at Harvard Observatory in Massachusetts (USA), discovered the relation between the luminosity and the period of the Cepheid variable stars, called Leavitt's Law. This provided astronomers with the invaluable first *astronomical candle*, i.e. an object with a precisely known pattern, period-luminosity in this case, which can be used to measure the distance to far away objects. This was the tool used some years later by Edwin Hubble, at Mount Wilson Observatory in California (USA), in order to obtain distances to spiral nebulae with the Hooker Telescope, the largest

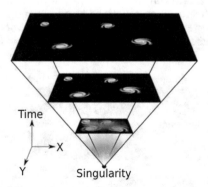

Fig. 3.3   Artist's illustration of the expansion of a slice of a small portion of the Universe. According to the Big Bang theory, the Universe expanded from an extremely dense and hot initial state, called singularity, in which the known laws of physics are not valid, and continues to expand today, carrying galaxies with it like spots on an inflating balloon. *Credit* Courtesy of Fredrik and Wikimedia Commons (Creat. Comm. 4.0 license)

Fig. 3.4 Vesto Slipher (1875—1969) in 1909. He was astronomer at Lowell Observatory in Flagstaff, Arizona (USA) from 1901 to 1954. In 1912 he observed for the first time the shift of the spectral lines of galaxies and, accordingly, computed their velocities with respect to the Earth. He found that most galaxies were in recession and published a table with the redshifts of 41, which included a dozen with large velocities from 800 to 1800 km/s. He provided therefore the first empirical basis for the expansion of the Universe. *Credit* Courtesy of Wikimedia Commons

telescope in the world from 1917 to 1949 and a technological feat at the time, acquired thanks to the financial support of a millionaire.

The second major contribution was made by Vesto Slipher (Fig. 3.4), at Lowell Observatory in Flagstaff, Arizona (USA), who observed for the first time the shift of the spectral lines of spiral nebulae. On September 17, 1912, pointing his telescope at our neighboring galaxy Andromeda he measured its spectral lines and observed they were blueshifted, leading him to the conclusion that Andromeda was approaching the Earth at a speed of about 300 km/s. In this way, he obtained the first radial velocity[3] of a spiral nebula ever. In addition, Slipher reported in 1914 the first observation of the rotation of such galaxies. In the meantime, he performed more measurements of

---

[3]The radial velocity denotes the speed with which the galaxy moves away from the Earth, or approaches it, in the case of a negative radial velocity. It is usually called just velocity.

spiral nebulae spectral lines and realized that large velocities up to 1100 km/s, most of them in recession, were a general property of those nebulae.

In 1914 Slipher presented the spectral lines corresponding to 15 spiral nebulae to the American Astronomical Society and received a standing ovation. All but three showed redshifts, indicating the galaxies were moving away from us; the other three showed blueshifts, indicating they were coming towards us, and they were also close in our galactic neighborhood. Thus, the blueshifted galaxies were approaching because they were gravitationally bounded to our Milky Way. By 1917 he published the results of 25 spiral nebulae—only four of them showing blueshifts—and a few years later he published a table with the redshifts of 41, which included a dozen with large velocities from 800 to 1800 km/s. That table was included in the book *The Mathematical Theory of Relativity*, by Arthur Eddington, published in 1923. Vesto Slipher provided therefore the first empirical basis for the expansion of the Universe, in which Eddington firmly believed.

Back to Lemaître, during his stay at MIT he visited Vesto Slipher at Lowell Observatory and Edwin Hubble at Mount Wilson. Both kindly passed him their respective tables of velocities and distances of the spiral galaxies. From the data it was easy to infer the recession of most galaxies with respect to our Milky Way, with velocities approximately proportional to their distances. Lemaître made a graph "velocity versus distances" showing this roughly linear relationship between them that years later was dubbed as Hubble's Law. This relation seemed to indicate that the Universe was expanding since the greater the distances, the faster the galaxies moved away from us.[4]

Lemaître published this work in 1927 in the article *A homogeneous Universe of constant mass and growing radius accounting for the radial velocity of extra-galactic nebulae* (in French) in the little known Belgian journal "Annales de la Société Scientifique de Bruxelles". As a consequence, his results passed unnoticed. It should be pointed out that Lemaître understood the expansion of the Universe in terms of the solutions of General Relativity—Einstein's theory of gravitation—for the case of a homogeneous Universe; that is, a Universe where all positions in space are equivalent. In fact, Lemaître and Friedmann had found, independently, that the solutions of General Relativity for a homogeneous Universe implied that it was not static but had to be either expanding or contracting.

Two years later, in 1929, Edwin Hubble (Fig. 3.5) presented his own graph "velocity versus distances" in the article *A relation between distance and radial velocity among extra-galactic nebulae*, using Slipher's velocity table of spiral

---

[4]This can easily be seen by inflating a balloon with some spots marked on its surface; the bigger the distance between the spots, the faster they move away from each other.

Fig. 3.5 Edwin Hubble (1889–1953) in 1931. He was astronomer at Mount Wilson Observatory, in California (USA). In 1925 he proved that the "spiral nebulae", previously thought to be clouds of dust and gas, were actually galaxies like the Milky Way located far away. This settled the so-called First Great Debate (whether or not the Milky Way was the whole Universe). On the right, one can see the 2.5 m diameter Hooker Telescope that Hubble utilized. It was the largest telescope in the world. Apart from the Hubble–Lemaître Law, Hubble's name is most widely recognized for the Hubble Space Telescope, which was named in his honor. *Credit* Courtesy of Wikimedia Commons and also of Andrew Dunn (right) (Creat. Comm. 2.0 license)

nebulae. These results were enlarged in a subsequent article in 1931 with his assistant Milton Humanson, who performed more measurements of galactic redshifts. However, in none of these articles did the authors discuss the interpretation of the galactic redshifts as a consequence of the expansion of the Universe, an expansion which was embodied in the solutions of General Relativity, as pointed out before. As a matter of fact, it is not clear to what extent Hubble believed in such an expansion; many astronomers at that time did not, and proposed alternative explanations for the galactic redshifts (including the so-called *de Sitter effect* that related the redshifts to the curvature of an static Universe).

The linear relation between radial velocity v and distance d was named Hubble´s Law shortly afterwards, but recently was renamed *Hubble-Lemaître Law* by the International Astronomical Union during its "IAU XXX General

Assembly" (Vienna, August 2018). This law, which is the basis for the model of the expanding Universe, states that at large scale the dominant motion of the Universe is dictated by the relation:

$$v = H_0 \, d, \tag{3.1}$$

where H_0 was called *Hubble's constant*. Its value is approximately[5] 70 km/s/Mpc. Hubble's estimate for H_0 was 500 km/s/Mpc, however, far away from the current value due to large inaccuracies in the distance measurements. Lemaître's estimate of 625 was even worse, although not that different from Hubble's since they had used essentially the same data. As a matter of fact, Hubble's constant H_0 is the present value of the Hubble parameter H, which gives the relative variation of the *scale factor* of the Universe with time: $H = \dot{a}/a$, where a is the scale factor that measures how much the Universe has expanded between two different epochs, and $\dot{a}$ is its time derivative indicating the rate of change of a with time.

As mentioned, Lemaître proposed in 1931 the rather rudimentary theory of the primeval atom. According to this theory, the Universe began with the explosion of that atom, but it still took two decades for a more elaborated version of that "explosion" to be called the *Big Bang*. Interestingly, this term was coined by the astronomer Fred Hoyle, who used it on a BBC radio program in 1949 to describe Lemaître's theory, which he did not believe. In fact, Hoyle was a proponent of the *Steady State* theory of the Universe, which he had elaborated in 1948 together with Thomas Gold and Hermann Bondi and was the alternative to Lemaître's theory at that time. Apparently, Hoyle totally rejected the idea that the Universe had a beginning, not so much the expansion of the Universe as such. He said, word for word: "*These theories were based on the hypothesis that all matter in the Universe was created in one Big Bang at a particular time in the remote past*". And he said that with an intonation that clearly showed he dismissed the idea as absurd, totally ridiculous.

After Lemaître, it was George Gamov who most supported the Big Bang theory, developing some of its implications. Among these was the study of the *primordial nucleosynthesis*, which he undertook together with his student Ralph Alpher. This is the synthesis of the lightest atomic nuclei, which took place between 10 s and 20 min after the beginning of the Universe. From these nuclei, the stable ones correspond to the isotopes ^2H,

---

[5]This means 70 km/s per million parsecs (Mpc), where the parsec is the standard unit for large astronomical distances, its value being: 1 pc = 3.2616 light-years, about 3.1×10^{13} km.

^{3}He, ^{4}He and ^{7}Li, of hydrogen (deuterium), helium and lithium. Remarkably, the Big Bang theory could explain both the formation and the observed abundances of ordinary hydrogen, deuterium and helium in the Universe. This fact alone gave a clear advantage to this theory over the rival one of the Steady State Universe.

But what finally catapulted the Big Bang theory and buried its rival was the discovery by Arno Penzias and Robert Wilson, in 1964, of the *Cosmic Microwave Background Radiation*, CMB, which permeates the entire Universe in all directions with a temperature of 2.7 K (−270.4 °C). This radiation was another prediction of the theory and was first proposed by Ralph Alpher and Robert Herman in 1948. It should have been released once the Universe cooled down sufficiently for protons and other light nuclei to be able to capture electrons and form atoms.

That period of formation of the primordial atoms is known as *era of recombination*, which is a misnomer because no "combination" era existed previously. It happened about 380,000 years after the Big Bang, when the temperature of the Universe dropped to a value of around 3000 K (2727 °C). Until that time, photons had been trapped in the plasma of charged particles, mostly electrons, protons and α particles, moving frantically between them, and it was the constitution of the electrically neutral atoms what allowed photons to free themselves. The possibility that in this plasma there was also primordial antimatter, in the form of positrons, antiprotons and $\overline{\alpha}$ antiparticles, is a subject of debate and even controversy, as will be discussed in Chap. 7. We see therefore that the appearance on stage of atoms (*and antiatoms?*) made the Universe transparent to light, to electromagnetic radiation. The photons that were released, about 1.61×10^9 per proton, are the ones which constitute the CMB radiation that we observe today, whose wavelength has increased by a factor of 1300 due to the expansion of the Universe.

After the discovery of Penzias and Wilson, the CMB radiation was analyzed by several experiments, like the BOOMERANG balloon experiment, launched in 1998 near the South Pole, and especially by the COBE, WMAP and Planck satellites, shown in Fig. 3.6. The first spacecraft, the *Cosmic Background Explorer* (COBE), was launched in 1989, and after four years operating it could confirm the expectations that the spectrum of the CMB radiation corresponds very accurately to a black body[6] at a temperature of 2.7 K. The *Wilkinson Microwave Anisotropy Probe* (WMAP) was next,

---

[6]A black body is an idealized physical body that absorbs all incident electromagnetic radiation. It also emits radiation and when in thermal equilibrium, at a constant temperature T, its spectrum of emission, which is a continuum without spectral lines, is determined by the temperature alone.

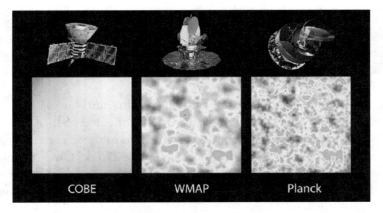

Fig. 3.6 Comparison of the anisotropies of the CMB radiation obtained from the data of the COBE, WMAP and Planck space missions, the first two from NASA and the latter from the European Space Agency (ESA) with contributions from NASA. The three panels show 10-square-degree patches of the full-sky maps created by these missions, where the successive improvement in the resolution of the images is remarkable. One can also see the satellites, which were specifically designed to measure the CMB radiation. *Credit* Courtesy of NASA, JPL-Caltech, and ESA

from 2001 until 2010, and the *Planck Space Observatory* followed. Launched in 2009 and deactivated in 2013, the Planck spacecraft obtained the best resolution full-sky maps of the CMB radiation, which were published in 2013, 2015 and 2018. These missions found a high degree of isotropy in the CMB radiation—almost identical values in all directions. Furthermore, the instruments in BOOMERANG, WMAP and Planck could also measure very small anisotropies, of order 10^{-5} (a few parts in 100,000). These anisotropies are of tremendous importance since they were the seeds that gave rise to the large structures present in the Universe (galaxies, galaxy clusters, and superclusters), among other properties. We will return to this issue in Sect. 3.4.1.

3.2 Antimatter, Dark Matter and Dark Energy: Key Features

First of all, we must clarify that in Astrophysics and Cosmology the term "matter" refers to anything that produces gravitational attraction and whose density dilutes with the expansion of the Universe in proportion to the increase of its volume, that is, in proportion to the cube of the scale factor a^3. This means that the density of any type of matter decreases as a^{-3} with the Universe's expansion. According to this, one finds that the matter in the Universe belongs to three different kinds: ordinary atomic matter from which atoms are made, antimatter and dark matter, which includes at least two species of neutrinos and antineutrinos. Atomic matter is often referred to as *baryonic matter*. The reason is that atoms owe their masses mainly to the baryons in their nuclei—protons and neutrons—since electrons are 1836 times lighter. Therefore, what is usually called baryonic matter consists actually of protons, neutrons and electrons in any possible configuration, such as atoms, ions, or plasmas. In the text that follows we will use as much as possible the more correct term "atomic matter" over "baryonic matter".

Antimatter

As was explained in the previous chapter, antimatter can be viewed as the reverse of ordinary matter since particles and their antiparticles have opposite properties, and when the two come into contact they annihilate each other, resulting in the release of radiation. Each elementary particle has an antiparticle partner, the two corresponding to the excitations or perturbations of the same quantum field. This is expressed formally by describing particles and antiparticles as the two solutions of the same equation; for example, the Dirac equation. Therefore, we understand the nature of antimatter and ordinary matter equally well since they are the two sides of the same coin.

A striking property of antimatter is that, to our knowledge, it contributes less, or much less, than 0.000001% to the total matter of the Universe as well as to its energy density (the exact amounts are unknown). Atomic matter, by contrast, contributes 15% to the total matter and 4.9% to the energy density. This asymmetry in the abundances of atomic matter versus antimatter gives rise to one of the great mysteries of Physics: Why is there so much atomic matter, in comparison with antimatter, in our Universe?

Dark Matter

Dark matter is a proposal to explain the invisible sources of mysterious gravitational fields which are abundant throughout the Universe, being particularly noticeable around spiral galaxies and galaxy clusters. Indeed, it holds galaxies together and triggers galaxy formation. Dark matter does not seem to emit or absorb any electromagnetic radiation at any frequency in the whole spectrum. Hence, it is not really dark but invisible, like a colorless transparent glass. Neutrinos and antineutrinos are part of the so-called *hot dark matter* because they travel at very high speeds, but provide a rather small contribution to the total dark matter, at most 1.5%.

Also, fairly large amounts of celestial bodies can add a small contribution to the total dark matter if our telescopes do not receive enough electromagnetic radiation from them, making their detection impossible. This contribution is dubbed *baryonic dark matter*. Examples include white

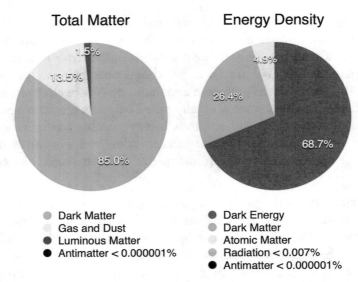

Fig. 3.7 Contributions of the components of the Universe to its total matter (left), and to its energy density (right), according to the Planck mission data analysis released in 2018. Dark matter dominates the total matter with a contribution of 85%, and adds about 26.4% to the energy density. Ordinary atomic matter represents 15% of the total matter, from which only 1.5% corresponds to bright luminous matter; the remaining 13.5% comes mainly from non-luminous gas and dust clouds. It only contributes 4.9% to the energy density. Dark energy dominates the energy density with a contribution of 68.7%. The contributions of radiation and antimatter (the latter not accounted for by the Planck data) are so minuscule that they cannot be seen in these charts

dwarf stars, neutron stars, brown dwarfs, planets, black holes and cold clouds of gas and dust.[7] Even so, the main components of the dark matter are as yet unknown, despite the fact that it dominates the total matter of the Universe with a contribution of 85% and it also adds about 26.4% to its energy density (Fig. 3.7).

Dark Energy

Dark energy is the name given to the unknown cause of the current accelerated expansion of the Universe. Its nature is also a mystery, although the cosmological constant Λ—a term in Einstein's equations of General Relativity—is a very strong candidate. This energy permeates the entire Universe in a uniform manner, dominating its energy density with a contribution of around 68.7%, but it does not add to the matter of the Universe because it

---

[7]White dwarf stars are described below in Sect. 3.5.1 while neutron stars and black holes are briefly introduced in next chapter, Sect. 4.6.2.

does not produce gravitational attraction (Fig. 3.7). Unlike matter and radiation, dark energy is essentially not diluted by the expansion of the Universe; this is precisely the reason behind its dominance at the present time.

3.3 Composition of the Universe

Besides antimatter, dark matter and dark energy, the other components of the Universe are atomic matter and radiation.

Atomic Matter
As noted above, atomic matter represents 15% of the total matter in the Universe and contributes a mere 4.9% to its energy density. However, bright luminous matter, like stars, active galactic nuclei (AGNs) and nebulae only accounts for 10% of the atomic matter; the remaining 90% being non-luminous and occurring mainly in the form of interstellar gas and dust clouds (Fig. 3.7). Luminous objects refer strictly to those seen in the visible optical range of the electromagnetic spectrum, i.e. at wavelengths perceived by the human eye, from about 380 to 740 nm. Visible objects, on the other hand, often refer to those objects that can be seen at some wavelengths, but not necessarily in the optical range. This is the case for the non-luminous atomic matter, which is not detected by our optical telescopes but is seen at other frequencies such as radio, infrared, ultraviolet, X-ray and γ-ray frequencies. In this respect, it should be noted that atomic matter always emits electromagnetic radiation, at least thermal radiation, as long as its temperature is above zero Kelvin (0 K).

Non-luminous atomic matter also comes in the form of stellar remnants—white dwarfs, neutron stars and black holes—as well as brown dwarfs, planets, and other small bodies. However, as said above, a fraction of these astronomical objects contributes to the dark matter instead because they are very difficult to detect. This especially applies when they are located in the halos of galaxies (more details in Sect. 3.4.2).

Radiation
Radiation consists of all relativistic particles, i.e. particles travelling at the maximum speed c or very close to it. At the present stage of the Universe, radiation consists mainly of photons. Curiously, although photons are the most abundant Standard Model particles in the Universe, their contribution to its energy density is only 0.005%. They do not contribute to the total matter because radiation does not qualify as matter due to the fact that its energy density decreases as a^{-4} with the expansion of the Universe.

Another component of radiation could be neutrinos (and antineutrinos). These are the only known particles that behaved as radiation in the early stages of the Universe and as dark matter later on. This has to do with the fact that, unlike photons, neutrinos have a very tiny mass, and this makes all the difference because massless particles are always part of radiation. Indeed, if the three species of neutrinos had no mass, then neutrinos and antineutrinos would add 0.0034% to the energy density of the Universe, so that the total contribution due to radiation would be 0.0084%. But this is not the case, since at least two of the three neutrino species are not relativistic and therefore they are not part of radiation but of hot dark matter instead.

The reason for this strange behavior is that neutrinos become non-relativistic when their thermal energy—$3/2\, k\, T_\nu$—falls below the rest mass energy—$m_\nu\, c^2$—where k is the *Boltzmann constant* and T_ν the temperature of the *Cosmic Neutrino Background Radiation*, and this is precisely the situation of at least two species of neutrinos in the Universe today, with $T_\nu \cong 1.95$ K. As a matter of fact, the exact values of the neutrino masses m_ν are unknown, as pointed out in the previous chapter, although rough estimates can be obtained from the known differences in their squared-masses and from the cosmological constraints on the total value of the three neutrino masses.

To summarize, the energy density of the observable Universe consists almost exclusively of dark energy (68.7%), dark matter (26.4%) and atomic matter (4.9%), in that order (Fig. 3.7). Radiation contributes less than 0.007% and the contribution of antimatter is expected to be negligible, probably much less than 0.000001%. It should be pointed out, however, that these contributions (except the one for antimatter) correspond to the latest Planck mission study, published in 2018, and they may therefore change in the future. As a matter of fact, in many articles and books one can see the outdated contributions given by the previous mission WMAP (roughly, 73% for dark energy, 23% for dark matter and 4% for ordinary atomic matter).

Another important consideration is that all these contributions have been obtained on the basis of the current Standard Model of Cosmology—*the concordance ΛCDM Model*—where Λ stands for the cosmological constant and CDM for cold dark matter. This model assumes that the Universe is isotropic and homogeneous at very large scales, as the astronomical observations (CMB, galaxy counting and others) seem to suggest. Even so, there is some controversy about these assumptions in the scientific literature.

3.4   Dark Matter: Invisible Source of Gravity

3.4.1   The Case for Dark Matter

The observation that our Universe contains large amounts of unseen matter dates back to the nineteenth century. In 1884 Lord Kelvin was among the first to provide an estimate of the number of dark bodies in the Milky Way, which was the only known galaxy at that time. As it turned out, before 1925 most astronomers believed that spiral nebulae and anything else in the skies were objects inside our own Milky Way, perhaps early stages of planetary systems. The work of Edwin Hubble measuring the distance to Andromeda (more than two million light years) put an end to many years of discussions on this subject—the so-called First Great Debate—that "officially" had started on April 26, 1920, but that actually was centuries old. In fact, amazingly, the eighteenth century philosopher Immanuel Kant had already suggested that the spiral nebulae were extragalactic "island universes".

Lord Kelvin's strategy was to assume that the stars in the Milky Way can be described as a gas of particles acting under the influence of gravity. Then, from the observed velocity of the stars he estimated the mass of the galaxy and he concluded that it was much larger than the total mass corresponding to the visible stars:

> It is nevertheless probable that there may be as many as 10^9 stars but many of them may be extinct and dark, and nine-tenths of them though not all dark may be not bright enough to be seen by us at their actual distances. [...] Many of our stars, perhaps a great majority of them, may be dark bodies.

The first astronomer who proposed the existence of large amounts of non-luminous unseen matter far beyond the Milky Way was Fritz Zwicky in 1933. While working at Caltech (California Institute of Technology, USA) investigating the Coma galaxy cluster—about 1000 galaxies held together by gravitation—he made an important discovery. He realized that the velocities of many individual galaxies were about 10 times higher than expected taking into account only the gravitational attraction of the visible galaxies, which was far too small for such fast orbits. At those speeds, the galaxies would have drifted apart, and the cluster would have been disintegrated. Therefore, something hidden from view generated a much stronger gravitational pull that kept the galaxies in place. Zwicky called this unseen matter "dunkle Materie" (dark matter, in German).[8]

Rotation Curves of Spiral Galaxies

The work of Zwicky about dark matter was not taken seriously by his colleagues at that time. In fact, it took about 40 years for the question of the existence of dark matter to reappear. This happened in 1970, when Vera Rubin (Fig. 3.8), Kent Ford, and independently Ken Freeman, provided further strong evidence by studying the rotation curves of stars in spiral galaxies. They found, similarly, that the luminous mass of the galaxies was much smaller than the mass needed to keep the individual stars in place orbiting the center of the galaxy. In other words, the observed gravitational strength and the gravitational pull of the visible stars (through optical telescopes) did not match, and the latter could not possibly prevent many stars from flying away from the galaxy.

Moreover, the rotation curves had a peculiar behavior since all the peripheral stars rotated at almost the same speed regardless of their distance to the galactic center. This flattening of the curve was very bizarre since, according to Newtonian gravitation, the orbital velocity v of a peripheral star should decline in inverse proportion to the square root of its distance r from the center of the galaxy, as given by the relation:

$$v = \sqrt{\frac{GM}{r}}, \tag{3.2}$$

where G is Newton's gravitational constant and M the mass of the galaxy enclosed by an imaginary sphere of the same radius r as the orbit of the star. Therefore, the stars at the periphery should turn slower than those closer to the galactic center. However, this formula also indicates that the velocity v of the stars can remain constant as the distance r varies, as long as the mass M of the sphere enclosed by the orbit grows at the same rate as r. In other words, if the ratio M/r remains constant, then the velocity v remains constant as well.

It should be noted that at the same time that Rubin and Ford investigated the rotation curves of spiral galaxies with optical telescopes, radio astronomers started using radio telescopes to study the 21 cm spectral line of atomic hydrogen (H-I) in nearby galaxies. In 1972, David Rogstad and Seth Shostak published the H-I rotation curves of five spiral galaxies showing

---

[8]Although born in Bulgaria in 1898, Fritz Zwicky was Swiss as well as a USA citizen. He spent most of his professional life at Caltech, where he made valuable contributions to the theory of supernovae and neutron stars, among other remarkable achievements.

Fig. 3.8 Vera Rubin (1928–2016) was mainly known for her work on rotation curves of stars in spiral galaxies, providing evidence for the existence of dark matter. This portrait of 1996 with some of her collection of globes is in the National Air and Space Museum of the Smithsonian Institution. Recently, in December 2019, the *Large Synoptic Survey Telescope* (LSST), an astronomical observatory currently under construction in Chile, was renamed as the *National Science Foundation Vera C. Rubin Observatory*. This is the first time that a national U.S. observatory has been named in honor of a woman astronomer. *Credit* Mark Godfrey. Courtesy of the Smithsonian Institution

that all five were very flat in the outer parts of their extended H-I disks (see Fig. 3.9). Some other measurements made by different research groups confirmed that in the galactic disks there are enormous amounts of hydrogen gas, also rotating in the shape of disks, but extending far beyond the stars. This allowed to broaden the sampling of rotation curves and, therefore, to increase the data of the total mass distribution in the galaxies.

The simplest explanation that fitted all the data was that there were huge, roughly spherical, halos of invisible dark matter, about 50 times more massive than the luminous stars, within which the galaxies were embedded (Fig. 3.10). This suggested that most galaxies must contain about five times

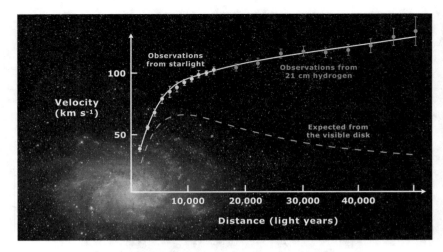

Fig. 3.9 Rotation curve of the spiral galaxy Messier 33—yellow and blue points with error bars—where one can appreciate the flattening with respect to the curve deduced from the distribution of the visible matter, depicted by the gray line. The maximum speed is located at a few thousands parsecs from the center. Then it should decrease in inverse proportion to the square root of the distance to the center, which is not the case. The discrepancy between the two curves can be accounted for by embedding the galaxy inside a huge roughly spherical halo of dark matter extending far beyond the galaxy. *Credit* Courtesy of Mario De Leo and Wikimedia Commons (Creat. Comm. 4.0 license)

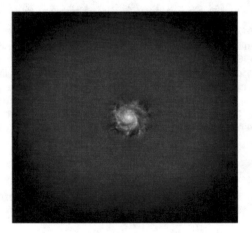

Fig. 3.10 Artist's illustration showing the dark matter halo around the luminous part of the spiral Pinwheel galaxy, similar to our Milky Way. Dark matter, represented in green, is however invisible—colorless and transparent—like pure water, and penetrates the entire galaxy as well. *Credit* Photo of the Pinwheel galaxy: Hubble Space Telescope, courtesy of NASA and ESA

more dark matter than the matter observed through electromagnetic radiation, taking into account that the giant non-luminous hydrogen clouds, visible on radio and X-rays, are nine times more massive than the luminous stars. In any case, around 1980 the scientific community had fully recognized that dark matter was a well-established hypothesis and, in fact, a major unsolved problem in Astrophysics, Particle Physics and Cosmology.

Other lines of evidence in favor of dark matter include: gravitational lensing, the analysis of the anisotropies of the CMB radiation, the formation and evolution of galaxies, and some misalignments between the center of mass of the visible matter and the center of mass of the gravitational lens during galactic collisions. Let us take a closer look at all these clues.

Gravitational Lensing

In the 1980s, a series of observations of gravitational lensing further supported the existence of dark matter. This phenomenon is a result of the curvature of light by gravitational fields. It turns out that photons feel the attraction of gravitational fields. As a consequence, when passing near a very massive body, like the Sun, light can bend enough for our instruments to detect the deflection. This effect, proposed by Einstein in 1911, encouraged several astronomers to embark on an expedition to Brazil in order to witness the total solar eclipse of October 10, 1912, and to measure the bending of the starlight near the Sun. Unfortunately, due to bad weather, the eclipse could not be accurately analyzed by any of the scientists. Anyway, Einstein's prediction of the light deflection by the Sun's gravitational field was flawed, as he himself realized. Three years later, in 1915, he presented the correct calculation as an outcome of his General Theory of Relativity.

A gravitational lens works under the same principles as an ordinary lens made of glass. The Sun, a galaxy or a galaxy cluster produces a gravitational lens for us when it is in the same line of sight between an astronomical body, such as a star, a galaxy or a quasar, and the Earth. Then, depending on the specific details, the resulting gravitational lensing can give rise to a variety of different optical effects in the object's image, such as distortion, magnification or multiple images like *Einstein crosses* and *Einstein rings* (see Fig. 3.11). Magnification is of particular interest because it allows researchers to study the details of early galaxies too far away to be seen with current telescopes.

Now, it is generally the case that the apparent origin of the gravitational lens, i.e. the observed matter, creates a gravitational field an order of magnitude weaker than needed to explain the observed lensing effects. In other words, the observed lensing effects indicate that the gravitational strength producing them corresponds to an amount of matter ten times bigger than the observed one. There are even cases where no matter is seen at all as the

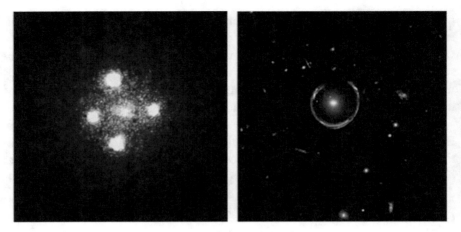

Fig. 3.11 On the left, the photograph shows the so-called Einstein Cross, four images of a very distant quasar which has been multiple-imaged by a relatively nearby galaxy acting as a gravitational lens. On the right, the gravity of a luminous red galaxy has gravitationally distorted the light from a much more distant blue galaxy. The lens alignment is so precise that the background galaxy is distorted into a nearly complete ring, known as Einstein ring. Strong gravitational lenses, like the ones in these two images captured by the Hubble Space Telescope (HST), allow to determine the mass and dark matter content of the foreground galaxy lenses. Observations using the HST have helped to greatly increase the number of Einstein rings known to astronomers. *Credit* Courtesy of NASA and ESA

origin of the gravitational lens. Hence, it is clear that gravitational lensing strongly supports the existence of dark matter and helps to understand its properties. Furthermore, images of gravitational lensing captured by the Hubble Space Telescope (Fig. 3.12) have been used to create maps of dark matter in galaxy clusters.

Anisotropies of the CMB Radiation

The tiny temperature anisotropies of the CMB radiation, of order 10^{-5}, enclose a great deal of information about the properties and composition of the Universe 380,000 years after the Big Bang, including its spatial curvature. These anisotropies correspond to the density perturbations or fluctuations of the ordinary matter that was present at that time and can be displayed in full-sky maps (Fig. 3.13). In addition, from the detailed analysis of the distribution and sizes of the CMB anisotropies, the full-sky maps can be decomposed into an *angular power spectrum* (Fig. 3.14). This spectrum bears many similarities with the decomposition of a given soundwave into its harmonics and contains a series of acoustic peaks at almost equal spacing but at different heights. It happens that any cosmological model predicts its own power spectrum, including the details of the peaks such as their positions

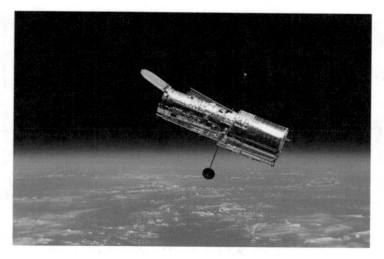

Fig. 3.12 The Hubble Space Telescope (HST), 13.2 m long, as seen from the Space Shuttle Discovery. It was built by NASA with contributions from the European Space Agency (ESA), and was launched on April 24, 1990. It stays in a stationary orbit around the Earth at an altitude of about 540 km completing a revolution every 95.42 min. Its four main instruments perform observations in the ultraviolet, visible, and near-infrared regions of the electromagnetic spectrum. The end of its mission is estimated to be between 2030 and 2040. *Credit* Courtesy of NASA

Fig. 3.13 Full-sky map of the temperature anisotropies of the CMB radiation obtained from the 2013 analysis of the Planck data. *Credit* The Planck Collaboration, ESA, and NASA

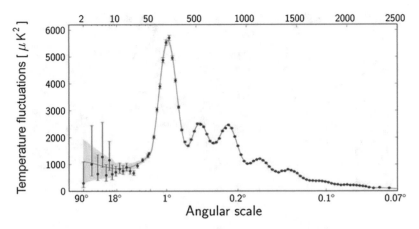

Fig. 3.14 Angular power spectrum of the full-sky map in Fig. 3.12. This graph shows the squares of temperature fluctuations (in micro Kelvins squared) in the CMB radiation at different angular scales on the sky. The first peak at the angular scale of 1 degree is exactly as predicted for a Universe with zero curvature. The red dots are the measurements made by Planck, showing seven acoustic peaks that are well fitted by a six-parameter ΛCDM model, represented by the green curve. At angular scales larger than six degrees, there is one data point that falls well outside the range of allowed models. This anomaly suggests that some aspects of the ΛCDM Model might be flawed. *Credit* Courtesy of ESA and the Planck Collaboration

and shapes, which are highly sensitive to small variations in the cosmological parameters. As a result, the actual power spectrum is used to restrict the cosmological parameters in a given model, or discard the model altogether, in order to match theory with observational data.

The position of the first peak of the CMB angular power spectrum (dubbed Doppler peak) is very sensitive to the spatial curvature of the Universe and its shape depends on the energy density of the total matter. This peak was first found by BOOMERANG, then by WMAP and finally by Planck, at the angular scale of 1 degree, exactly as predicted for a Universe with zero curvature (within 1% error). The height of the second peak determines the fraction of the atomic matter while the third peak is very sensitive to dark matter. In short, by analyzing the peaks of the CMB power spectrum one can obtain all the information about the composition of the Universe and its spatial curvature.

The fact that the spatial curvature of the Universe is almost zero is also expressed by saying that the Universe is nearly spatially flat. This means that the average density of the Universe, ρ, is almost the critical one that leads to zero curvature and is given by:

$$\rho_c = \frac{3H^2}{8\pi G},$$ (3.3)

where H is the Hubble parameter and G is Newton's constant. The critical density ρ_c varies with time because H is time dependent, and its current value is, approximately, $\rho_{c,0} = 9 \times 10^{-27}$ kg/m^3. We will come back to the curvature of the Universe in Sect. 3.5.1, where we will discuss its dependence with respect to dark energy.

As mentioned earlier, the 2018 analysis released by the Planck collaboration yielded a value of about 31.3% for the energy density of the total matter and around 4.9% for that of the atomic matter. Therefore, the observed CMB power spectrum provides powerful evidence in support of dark matter, with an energy density of about 26.4%.

Structure Formation

According to the dark matter hypothesis, the density perturbations or fluctuations of dark matter appeared first and evolved differently than the ones of atomic matter. The effects of dark matter on the CMB photons were in fact negligible, only by gravitational interactions. But on the atomic matter the effects of dark matter were crucial and played a major role in generating the density perturbations that led to the observed structure formation: galaxies, galaxy clusters, superclusters, filaments, lumps, as well as huge regions devoid of any galaxies, called voids, created by the configurations of the filaments and lumps. The point is that, without the aid of dark matter, a highly homogeneous early Universe dominated by radiation—photons and neutrinos—would not have had the time to evolve into the Universe we see today.

The reason is that atomic matter was interacting strongly with the photons present there and this washed out its density fluctuations. By contrast, the fluctuations of dark matter density were unaffected by radiation, so they could grow first and were able to attract atomic matter, speeding up the creation of density perturbations of the latter and their subsequent condensation into structure. As a matter of fact, computer simulations show that starting with a perfectly homogeneous Universe of only dark matter, gravity would produce at present filaments and lumps very similar to those where most galaxies are located. This effect can be appreciated in the image shown in Fig. 3.15, that represents a region of the only-dark-matter Universe of about 1700 million light years in size, where brighter spots represent more dense regions of dark matter where galaxies would cluster. *This strongly suggests that it is dark matter that shapes the large-scale structure of the Universe.*

The Bullet Cluster

The Bullet Cluster, two clusters of galaxies in an advanced state of collision, provides the best example known so far of the aforementioned misalignments between the center of mass of the visible matter and the center of mass of the gravitational lensing. It turns out that the collision affects mainly the giant gas clouds, visible in X-rays, since they interact electromagnetically with each other. As a result, they lost speed and lagged behind the stars, i.e. the luminous part of the galaxies, that were barely affected by the collision and passed right through. However, the lensing effects of background objects show that the lensing is strongest in the two regions around the luminous component, even though the mass of the gas clouds is about nine times the mass of the luminous stars. Therefore, this clearly indicates that most of the mass in the Bullet Cluster is concentrated in two regions of dark matter around the luminous part of the galaxies, and therefore this dark matter crossed the gas clouds during the collision (see Fig. 3.16).

Nevertheless, in spite of all the accumulated evidence supporting the existence of dark matter, there are research groups that consider the possibility that General Relativity is incorrect and must be modified to correctly calculate spiral galaxy rotation curves and gravitational lensing, among other

Fig. 3.15   Computer simulation of the distribution of dark matter in a region of the Universe—about 1700 million light years in size—at the present time. By comparing such simulated data to actual large surveys, the physical processes that underlie the build-up of real galaxies can be clarified. *Credit* The Millennium Simulation Project, Springel et al. (2005)

Fig. 3.16   The Bullet Cluster, a merger of two clusters of galaxies. The X-ray emission of the gas clouds of atomic matter is shown in pink, while the blue "clouds" represents the mass distribution calculated from gravitational lens effects. This misalignment strongly suggests the existence of dark matter separated from the bulk atomic matter as a result of the collision. *Credit* Courtesy of NASA and *Chandra X-Ray Center* (*CXC*)

effects. However, it should be noted that the Bullet Cluster provides a real challenge for the modified theories of gravity that disregard dark matter. Simply stated, dark matter models can easily explain the phenomena observed in the Bullet Cluster, while for the modified theories of gravity this is a much harder task.

3.4.2 Dark Matter Candidates

At present there are many candidates that could contribute to the dark matter in various proportions. A key factor in discerning between the different possibilities is the *free streaming length* of a particle, which indicates the distance traveled by the particle when it propagates through a medium without being dispersed. This length is related to velocity, and is used to classify dark matter as cold or hot; the latter being quasi-relativistic particles traveling at near the speed limit c. Now, for a number of reasons based on observational data, such as those mentioned in the previous paragraphs, it turns out that *most dark matter has to be cold*, that is, much slower than c. This is also necessary for the emergence of structures—galaxies, galaxy clusters and superclusters—in this hierarchical order, by gradual accumulation of their components, according to observation. Even so, as pointed out before, at least two of the three species of neutrinos are part of the hot dark matter and they contribute between 0.5 and 1.5% to the total dark matter.

Another piece of information comes from the observed abundances of hydrogen, deuterium and helium in the Universe, which are in reasonable agreement with the predictions[9] of the Big Bang nucleosynthesis and require that atomic matter constitutes between 4 and 5% of the energy density of the Universe. Therefore, the bulk of dark matter cannot be made of atoms because if there were more atoms than observed, then the predictions of the Big Bang nucleosynthesis would have proven wrong. Even so, as explained earlier, astronomical objects, such as stellar remnants, brown dwarfs and planets, as well as clouds of cold gas and dust, can also add a small contribution to the total dark matter, called baryonic dark matter. When these objects are in the halos of the galaxies, far from the luminous stars, they are referred to as *Massive Compact Halo Objects* (MACHOs). They are very hard to detect, but despite this difficulty astronomical searches for gravitational microlensing in the Milky Way have further confirmed that these objects cannot make up more than a small fraction of the dark matter.

We conclude that dark matter appears to be dominated by one or more unknown particles beyond the Standard Model of Particle Physics, like the ones mentioned at the end of the previous chapter (*axions* and WIMPS such as *neutralinos* and *Kaluza Klein particles*). But there are other possibilities. For example, hypothetical extended field configurations which are not made out of particles but are predicted in some theories, such as *cosmic strings, domain walls and monopoles*, could also add to the total dark matter. Another interesting dark matter candidates, which are gaining increasing acceptance, are *primordial black holes*, which could have resulted from enormous energy fluctuations in the first instants after the Big Bang. Some researchers even propose that primordial black holes alone may constitute almost the entire cold dark matter.

There are also several dark matter models grouped into a class called *ultra-light dark matter*, in which dark matter consists of very light bosons with masses ranging between 10^{-22} and 1 eV. These models have the interesting property of building a *Bose–Einstein condensate*[10] or a *superfluid* inside the

---

[9]The agreement between the expected and the observed abundances is less precise in the case of lithium.

[10]Bose–Einstein condensates and superfluidity are macroscopic quantum mechanical phenomena that can manifest themselves for boson particles, including atoms, at very low (cryogenic) temperatures. All the particles of the Bose–Einstein condensate behave as a single quantum state, described by the same quantum wavefunction. Superfluids have zero viscosity, which is why they form vortices when stirred up that can rotate indefinitely.

galaxies whereas outside of them there are just cold dark matter particles. They present a rich phenomenology with astrophysical implications amenable to being tested by future observations, including vortices and quantum mechanical interference.

Last but not least, a very intriguing possible source of dark matter comes from the framework of *extra dimensions*, from the so-called *brane-worlds*, where the term "brane" is a generalization of the two-dimensional membranes for more than two dimensions. The proposal is that our Universe could be embedded in a Cosmos with more than three spatial dimensions, and along one (or more) of the extra dimensions other universes parallel to ours could be located (Fig. 3.17). A crucial assumption is that, whereas the matter of every universe would remain confined within its own boundaries, the corresponding gravitational fields would propagate through the extra dimensions reaching the other universes as well. Consequently, the matter of all nearby universes would become aligned with each other along the extra dimensions and the collective gravitational fields would produce an *apparent dark matter* in each other's universes; in particular in ours. Therefore, in the brane-world scenario the bulk dark matter could simply consist of matter of diverse nature confined in other universes in our vicinity. Curiously, the idea that our

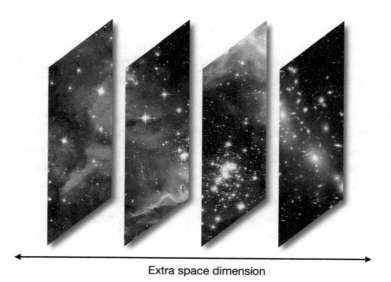

Extra space dimension

Fig. 3.17 The illustration shows a brane world consisting of (at least) four branes, which are parallel universes along an extra space dimension. In this example, the four branes are three-dimensional universes similar to ours with atoms and stars. However, in most brane worlds put forward so far, only one brane contains a Universe like ours, while the other branes have completely different particles and fields, and no atoms

Universe is embedded in a higher dimensional Cosmos with parallel universes is very old, especially in Eastern beliefs.

3.4.3 Dark Matter Detection

The possibility that dark matter is composed primarily of unknown subatomic particles opened new avenues in experimental Particle Physics aiming for their detection. For if dark matter is made up of subatomic particles, then they should fill the space around us like a fog of invisible particles; it is even suggested that millions or thousands of millions of such particles should be crossing us and the Earth per cm^2 every second. The many experiments designed to detect dark matter particles can be classified into three categories according to whether they are based on *direct detection, indirect detection,* or *particle production*. So far, none of these experiments have succeeded in providing a conclusive evidence of dark matter detection.

Experiments for the direct detection of dark matter, which started in the 1980s, aim to observe collisions of dark matter particles against the atomic nuclei of a material, which can be a very low temperature crystal, such as germanium, or a noble gas, like xenon or argon. The collisions would produce very slight recoils of the nuclei that could be measured by various techniques. These experiments need to be shielded from the cosmic ray background particles, especially muons, which could yield similar effects and would therefore interfere with the expected signals. For this reason, the experiments are installed deep underground in mines, or inside tunnels under very massive mountains.

Very recently, in June 2020 the XENON collaboration, hunting for dark matter since 2006, has reported significant signals pointing towards the existence of solar axions in the XENON1T experiment. However, those results can also be explained by β decays of tritium (^3H) contaminating the detector, located at the Gran Sasso National Laboratory, below the Gran Sasso mountain (Italy), which is filled with 3.2 tons of ultra pure liquefied xenon. Moreover, as pointed out in the previous chapter, the data seem to disagree with bounds of axions from the cooling of white dwarfs.

Indirect detection experiments look for signals of annihilation or decay of dark matter particles in outer space coming from regions of expected high dark matter density, like the center of the Sun or the center of the Milky Way. The hope is that these processes may result in known by-products, like neutrinos (and/or antineutrinos), γ rays, antiprotons, or positrons, leading to an excessive amount of such particles, which could be detected by a variety of instruments. Now, it turns out that already for many years, such excesses are seen, especially by space-based detectors aboard satellites or mounted on the International Space Station. In this respect, it should be underlined that the PAMELA experiment, since 2008, and then the AMS experiment have clearly detected a substantial excess of positrons (and a less substantial excess of antiprotons), but these antiparticles could also be produced by pulsars and black holes. We will come back to this subject in Chaps. 4 and 5.

Finally, dark matter particles might be produced by collisions in particle accelerators, like any other particles. However, the most powerful particle accelerator to date, the Large Hadron Collider (LHC) at CERN, has been searching for dark matter signals and has found no indication of their production, such as large amounts of missing energy in certain collision products.

3.5 Dark Energy: The Accelerating Universe

3.5.1 The Accelerated Expansion of the Universe

1998 will be remembered as the year in which our understanding of the Universe experienced yet another twist. Two international groups of observational cosmologists made a most amazing discovery: the rate of expansion of the Universe is accelerating rather than decelerating, contrary to the general expectation. These groups were the *Supernova Cosmology Project* team, led by Saul Perlmutter at the Lawrence Berkeley Laboratory (USA); and the *High Z Supernova Search Team*, led by Brian Schmidt, from Harvard University (USA) and the Australian National University at Weston Creek (Fig. 3.18). Both teams had been studying, for several years, the expansion history of the Universe using *type Ia supernovae*; in particular, they were striving to determine the deceleration experienced by this expansion since about 7000 million years ago.

What they found instead was that the expansion rate of the Universe had started accelerating around 5000 million years ago, as if under the influence of a repulsive force. But this could not possibly be caused by the known ingredients of the Universe—matter and radiation—since they produce gravitational attraction rather than repulsion. It is precisely the attraction of everything towards everything else, due to gravity, what made scientists believe that the expansion of the Universe had to slow down. And indeed, it was slowing down, but at earlier times; then something surprisingly turned the deceleration into acceleration and the expansion of the Universe began to speed up.

Fig. 3.18 Saul Perlmutter (left), Brian Schmidt (center) and Adam Riess (right) were awarded the 2011 Nobel Prize in Physics for the discovery in 1998 of the accelerating expansion of the Universe through observations of distant supernovae. Some years earlier, in 2007, Saul Perlmutter and Brian Schmidt received the Gruber Cosmology Prize together with their teams: the *Supernova Cosmology Project* and the *High-Z Supernova Search Team*, respectively. *Credit Photo* Ulla Montan © The Nobel Foundation

Type Ia supernovae (Fig. 3.19) are much better astronomical candles than the Cepheid variable stars for determining the distances to far away galaxies. These supernovae are thermonuclear explosions of white dwarf stars when their masses reach about 1.44 solar masses—*the Chandrasekhar limit*—producing an intrinsic brightness that follows a well-known light-curve pattern. It follows then that the distance to a galaxy hosting a type Ia supernova can be deduced simply by analyzing its apparent luminosity. The difference with the Cepheid stars is that supernova explosions are much more powerful; they are as bright as the entire galaxy that harbors them before they fade away over several weeks or even months. In fact, while Cepheid stars can be seen up to distances of a few million light years, supernovae can be seen at distances of thousands of millions of light years.

White Dwarf Stars in a Nutshell

White dwarfs are the final stage of small and medium size stars—until 10 solar masses—after exhausting their nuclear fuel and expelling their outer layers. Their gravitational collapse is stopped by the *electron degeneracy pressure*, which is a very technical effect due to the quantum mechanical *Pauli Exclusion Principle*. Most white dwarfs are composed of carbon and oxygen, have a mass comparable to that of the Sun and a size similar to Earth. Often, they are accompanied by an ordinary star, forming a binary system. If the two stars are

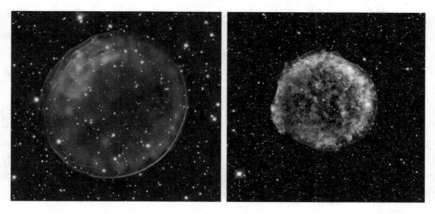

Fig. 3.19 Type Ia supernovae are thermonuclear explosions of white dwarf stars when their masses reach the Chandrasekhar limit, named after the scientist who computed it. On the left, one can see the image of the remnant of the type Ia supernova 0509–67.5, made using data from the Hubble Space Telescope and the Chandra X-ray Observatory. On the right, the remnant of the type Ia supernova SN 1572 that Tycho Brahe observed in 1572, 450 years ago. This composite X-ray and infrared image shows the expanding cloud of gas. *Credit* NASA and ESA (left); NASA and Max Planck Institute for Astronomy (MPIA) (right)

close enough, the white dwarf continuously draws matter from its companion until its mass reaches about 1.44 solar masses—the Chandrasekhar limit—giving rise to a type Ia supernova explosion, whose luminosity depends on the exact composition of the white dwarf. However, the corresponding "light-curves" are well understood, which is why type Ia supernovae can be used as standard candles.

The work of the two groups consisted of measuring the redshifts of galaxies hosting type Ia supernovae at different distances, using slightly different techniques to analyze the luminosity given by the light-curves. Let us remember that the Doppler redshift z of the spectral lines of a galaxy indicates its velocity of recession at the precise instant when the light was emitted. However, in an expanding Universe, General Relativity provides an interpretation for the redshifts different than that. Namely, the increase in the wavelength of light from a distant galaxy, $(\lambda_0 - \lambda(t))$, where $\lambda(t)$ corresponds to the light emitted by the galaxy at a given time t and λ_0 to the light received today in our telescopes, is mainly due to the expansion of the space between the galaxy and the Earth during the journey of the light.

The researchers were collecting data provided by the Hubble Space Telescope (Fig. 3.12) to obtain new velocity-distance graphs, like those of Lemaître and Hubble, with the advantage that they had access to much more distant galaxies, up to about 10,000 million light years. Now, taking into account that light propagates through space at the maximum speed c, it follows that looking at far away celestial objects implies looking at them as they appeared far back in time. In addition, due to the expansion of the space, the distance between the objects and the Earth gets stretched[11] while the light is traveling and, consequently, the objects look fainter than if no such expansion had taken place. Therefore, the velocity-distance graphs reveal the expansion history of the Universe and can be used to obtain the scale factor $a(t)$ as a function of the *cosmic time* t. This refers to the time measured by a clock that moves with the expansion of the Universe but without any peculiar velocity of its own.

The Scale Factor and the Cosmological Redshift

The meaning of the scale factor $a(t)$ is easy to grasp: if two far distant galaxies are separated by a distance d_0 at the current time t_0, then their distance $d(t)$ at any other earlier or later (cosmic) time t is given by

$$d(t) = a(t)d_0, \tag{3.4}$$

where at present the scale factor is set to 1, that is $a(t_0) = 1$. Hence, $a(t) < 1$ for past times, and $a(t) > 1$ for future times. This relation also holds for the wavelengths $\lambda(t)$ and λ_0, where $\lambda(t)$ corresponds to the light emitted by one of the galaxies and λ_0 corresponds to the stretched light arrived at the other one:

$$\lambda(t) = a(t)\lambda_0, \quad \text{with } a(t) < 1. \tag{3.5}$$

The relative velocity $v(t)$ between the two galaxies, at a given time t, is the rate of change of their distance at that time:

$$v(t) = \dot{d}(t) = \dot{a}(t)d_0, \tag{3.6}$$

where the upper dots mean time derivative and indicate the rate of change with time. Now, the relative acceleration of the galaxies at a given time t is given by the rate of change of the velocity at that time:

$$\dot{v}(t) = \ddot{d}(t) = \ddot{a}(t)d_0. \tag{3.7}$$

---

[11]This stretching causes our cosmological horizon - the furthest distance we can see - to be around 46,000 million light years, even though the age of the Universe is "only" about 13,800 million years.

Therefore, the acceleration of the expansion of the Universe is encoded in $\ddot{a}(t)$, the second derivative of the scale factor with respect to the cosmic time. The so-called *cosmological redshift* z is defined, like in the Doppler effect, as the relative increase of the wavelength:

$$z = (\lambda_0 - \lambda(t))/\lambda(t). \tag{3.8}$$

It can be expressed in terms of the scale factor using $\lambda(t) = a(t)\lambda_0$. One gets:

$$z = 1/a(t) - 1 \quad \text{and} \quad a(t) = 1/(z+1). \tag{3.9}$$

This implies that by measuring the redshift z of the light of a far away galaxy one obtains the value of the scale factor that tell us how much the Universe has expanded since the light was emitted.

Type Ia Supernovae Results

The two groups reported their findings in 1998. The *Supernova Cosmology Project* team presented the results of 42 supernovae at high redshift and 18 at low (Fig. 3.20). The highest redshift was $z = 0.83$, which corresponds to a supernova explosion that took place about 7000 million years ago (half the age of the Universe). The *High Z Supernova Search Team* presented the results of 14 supernovae at high redshift, and 34 supernovae at low redshift. In both experiments the high-redshift supernovae appeared between 10 and 15% fainter (therefore farther) than expected, and the observations could easily be explained if the rate of the expansion of the space had started accelerating around 5000 million years ago, when the Universe was some 8800 million years old. Furthermore, the two teams realized that the cosmological constant Λ in Einstein's General Relativity could be behind the "antigravity" repulsion causing the acceleration. Accordingly, they both argued that their respective results clearly pointed towards the existence of a positive Λ, whose energy density dominated the Universe for the last 5000 million years. We will come back to this subject in Sect. 3.5.2, where we will have a closer look at the cosmological constant Λ.

In no time, other cosmologists worldwide proposed different explanations for the observations of the Berkeley and Harvard groups, besides the cosmological constant Λ (including the possibility that there was no such a cosmic acceleration at all). Moreover, the unknown cause for the accelerated expansion of the Universe was called *dark energy*, Λ being only one of its possible candidates, although the preferred one. This name is not the best choice, however, for three reasons. First, dark energy is not dark but invisible, i.e. colorless transparent, like dark matter. Second, this name seems to indicate

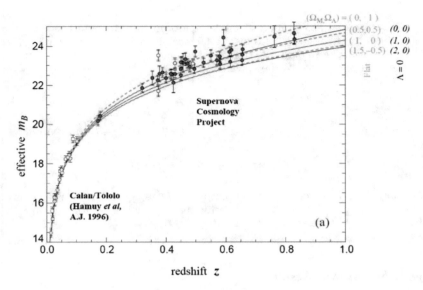

Fig. 3.20 Velocity-distance diagram presented by the "Supernova Cosmology Project" showing the results for 42 type Ia supernovae at high redshift (red dots) and 18 at low redshift (yellow dots). The distances, in vertical, are expressed in terms of a parameter related to luminosity. Several diagrams are depicted in blue and black showing the predictions for different values of the contributions of the total matter and of the cosmological constant Λ to the energy density of the Universe. The supernovae results favor the Universe with dominant Λ. *Credit* The Astrophysical Journal, 517, 565 (1999)

that this energy has a connection with dark matter, which is not necessarily the case. And third, it even suggests that dark energy is the energetic content of dark matter, using Einstein's formula for conversion between mass and energy.

At present, more than 20 years later, the so-called *acceleration of the Universe*, implying the existence of dark energy, has been verified using a wider sampling of supernova data, some as distant as 12,000 million light years, and also by other independent methods. In fact, the discovery of the acceleration of the Universe using supernovae was so successful that it received the Nobel Prize in Physics in 2011. The cosmological constant Λ is still the best candidate for being the dark energy, as it was from the very beginning. This is reflected in the fact that the Cosmological Standard Model is based on the assumption that the main components of the Universe are cold dark matter and Λ, which is why it is called the ΛCDM model, as mentioned before. Although this model is not free of controversy and under-lying problems, these have not been resolved by any of the alternative dark energy candidates either.

Age of the Universe

The accelerated expansion enlarges the size of the Universe, obviously, leaving an imprint on gravitational lensing and slowing down the rate of structure formation—galaxies, then galaxy clusters and finally superclusters. Less obvious is that the age of the Universe is also affected by that acceleration. As a matter of fact, using the General Relativity equations with no Λ or other type of dark energy, the age of the Universe results to be well below 12,000 million years. However, using astrophysical techniques, the age of the oldest stars is estimated to be somewhere between 12,000 and 15,000 million years! Therefore, according to General Relativity, without dark energy the Universe should be much younger than the oldest stars! This paradox was finally settled in 1998 thanks to the discovery of the acceleration of the Universe.

Baryon Acoustic Oscillations

There is another method to track the influence of dark energy, based on the analysis of the baryon acoustic oscillations (BAO) in the early Universe, that also points towards the existence of dark energy. These are fluctuations in the density of the baryonic (atomic) matter caused by acoustic pressure waves (like sound waves) in the primordial plasma before the recombination; that is before the formation of the first atoms about 380,000 years after the Big Bang. As it turns out, the details of those fluctuations depend on the components of the Universe that were present at that epoch, being very sensitive to the existence of dark energy. But those fluctuations of matter density are visible today, at large scales, in the form of wave patterns that can be measured and analyzed. Indeed, one finds that the spheres produced by the acoustic pressure waves have a radius of 490 million light years in the present Universe. Hence, BAO wave patterns provide a *standard ruler* for length scales in astronomical observations.

Anisotropies of the CMB Radiation

The anisotropies of the CMB radiation, translated into the angular power spectrum, provide further compelling evidence for the existence of dark energy, since they also depend on the composition of the Universe, as was explained earlier. In particular, the position of the first peak of the power spectrum revealed that the spatial curvature is almost zero, that is, the Universe is nearly spatially flat, i.e. on the verge of being closed and open (Fig. 3.21). However, other astrophysical methods had previously shown that the contribution of the total matter—atomic and dark matter—to the average energy density of the Universe, ρ, is much less than the critical one, ρ_c, that leads to a flat geometry, given by (3.3). Indeed, already from the data of the BOOMERANG experiment, it became clear that some other components of the Universe were missing. Amazingly, the dark energy contribution obtained from the observed acceleration of the Universe fitted exactly into this puzzle, providing the necessary energy density that finally resulted in an almost flat Universe.

Dark Energy Density

The density of dark energy is really tiny, about 6×10^{-27} kg/m^3 (where one uses the Einstein conversion formula between mass and energy), but still is the dominant component of the average density of the Universe, $\rho_{c,0}$, approximately 9×10^{-27} kg/m^3, as shown above. This density is much smaller than the density of both atomic and dark matter within galaxies. In fact, dark energy contributes to the mass of the Earth only 7 mg, filling its volume uniformly. Because of this lightness and because it does not interact with matter or radiation, dark energy is unlikely to be detectable in laboratory experiments. Yet, dark energy dominates the density of the Universe because

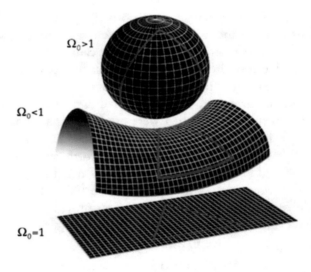

Fig. 3.21 The three possible spatial geometries of the Universe: closed, open and flat, from top to bottom, depending on the *density parameter* Ω_0, defined as the ratio: $\Omega_0 = \rho/\rho_c$. A closed universe is of finite size since traveling in a given direction will lead back to the starting point, and within it parallel light rays ultimately converge. The open and flat universes are infinite as traveling in a given direction will never lead back to the same point. However, in a flat universe parallel light rays remain always parallel whereas in an open universe they will eventually diverge. *Credit* Courtesy of NASA and WMAP Science Team

it is uniform throughout the space, which is essentially empty, with enormous spans between the galaxies and even larger spans between galaxy clusters and superclusters.

3.5.2 Dark Energy Candidates

The Cosmological Constant Λ

The cosmological constant Λ in Einstein's equations of General Relativity—the preferred candidate for the dark energy—represents the energy of empty space, i.e. pure space devoid of any matter and radiation, which is why it is referred to as the *vacuum energy*. It has a very intricate, even embarrassing, story. To start with, Einstein did not introduce this constant Λ in the first version of his theory, presented in 1915, as seen in Chap. 2, Eq. (2.4). He did so two years later for "aesthetic" reasons in an attempt to obtain a solution to his equations yielding a static Universe with no expansion or contraction. These equations, with the Λ term, read:

$$ R_{\mu\nu} - \frac{1}{2} R \, g_{\mu\nu} = \frac{8\pi G}{c^4} \, T_{\mu\nu} + \Lambda \, g_{\mu\nu}, \qquad (3.10) $$

where G is Newton's constant and c the speed limit. As explained in Sect. 2.4.5, the left hand side of these equations has to do with the geometry of the spacetime—Einstein's explanation for the "apparent" gravitational forces—and the right hand side is related to the specific matter-energy configuration under consideration, which is the source of such forces. By contrast, the cosmological constant Λ represents a constant energy that does not depend on any configuration.

As it turned out, Einstein failed miserably because the new solutions adding Λ were unstable; that is, they fixed "his" problem, although for a very short time. Moreover, once he became convinced that the Universe was really expanding, he declared that the introduction of Λ into his equations was a major blunder. Thus, *Einstein rejected the cosmological constant Λ in the same way that he had previously rejected the idea of an expanding Universe.*

In principle, in (3.10) Λ can take any value, including zero. If it is positive, $\Lambda > 0$, the solutions can actually lead to an accelerated expanding Universe depending on the details of the matter-energy configuration encoded in the stress-energy tensor $T_{\mu\nu}$. To be more precise, a positive Λ behaves like a repulsive force because it increases the distance between two celestial bodies, accelerating their mutual recession velocity by creating more and more space between them. Should Λ be negative, on the contrary, then it would join gravity behaving like an attractive force. Hence, in the case $\Lambda > 0$ the "repulsive forces" created by Λ had to overcome the attractive gravitational forces due to matter and radiation in order for the space expansion to start accelerating. Let us have a closer look at this.

The cosmological constant Λ has highly counterintuitive properties. One of them is that it permeates the entire space uniformly but *its density does not dilute with the expansion of the Universe*, in sharp contrast with both the matter and radiation densities, which decrease in inverse proportion to the scale factor, as a^{-3} and a^{-4}, respectively. This is precisely the reason why the Universe's expansion started accelerating around 5000 million years ago, when the Universe was some 8800 million years old. In simple terms, this happened because before that time the combined contributions of matter and radiation to the energy density were greater than the contribution of Λ. But as time was passing, matter and radiation were diluting with the expansion of space, until their combined energy density dropped below the immutable, constant energy density of Λ. Then the "repulsive forces" overpowered the attractive gravitational forces and, as a consequence, the expansion of the Universe began to accelerate.[12]

An important remark is that the meaning of "vacuum" is very different in General Relativity than in Particle Physics, where the vacuum in the sense of nothingness simply does not exist. As was explained in the previous chapter, the so-called vacuum in Quantum Field Theory is an idealized empty state without any particles,[13] but it is full with quantum fields that permeate the whole Universe, much like the cosmological constant. Moreover, the quantum fields generate *quantum fluctuations* that add to the energy of the "empty space", apart from other contributions courtesy of symmetry breakings and other technicalities abundant in the Standard Model of Particle Physics.

One might think that this situation is most welcome, since at least Quantum Field Theory provide good reasons for the existence of some sort of vacuum energy Λ. However, it turns out that the natural value of Λ, as deduced from the breaking of the electroweak symmetry, is at least 60 orders of magnitude (10^{60}) larger than the value obtained from the observed acceleration of the Universe. And the numbers become even worse if one takes into

---

[12]The successive dominant contributions to the energy density of the Universe define three cosmic eras in Cosmology. The first one was the radiation era, from the Big Bang until about 50,000 years later, when matter took over. Then there was the matter era, which lasted until some 5000 million years ago, giving way to the present dark energy era, also called "vacuum" dominated era.

[13]In the real Universe there are no places, not even tiny ones, without any particles because low energy photons and neutrinos and antineutrinos (and gravitons, if they happen to exist) permeate the whole "empty" space, traveling in all directions.

account common lore arguments from quantum gravity; then the discrepancy raises up to 120 orders of magnitude. Therefore, *there is something flawed in the identification of* Λ *with the vacuum energy of the quantum fields.*

Other Proposals

Besides the cosmological constant Λ, several other proposals have been made to explain the origin of the observed acceleration of the Universe's expansion and the nature of the dark energy. These proposals can be classified into two large classes, depending on whether the acceleration of the Universe is considered a real phenomenon, or not. In addition, the first class contains two subclasses, depending on whether dark energy is considered as some kind of substance, i.e. another component of the Universe, or not. If so, it might dilute with the expansion of the Universe, although in a much smaller proportion than that of matter and radiation.

In the models where dark energy is a substance, this might be very similar to Λ, a kind of vacuum energy although varying over space and time; this is the case for the *running vacuum models*. Alternatively, dark energy could result from the energy of unknown quantum fields which can be appropriately incorporated into the Standard Model. For example, the *quintessence models* assume the existence of a scalar field[14] that varies very slowly in space and time and whose energy produces the acceleration of the Universe. There are also the *interacting models*, in which dark matter and dark energy have a common origin since they are considered as two features of the same physical substance. Now, these models use modified theories of gravity correcting General Relativity, which often predict the speed of gravitational waves to be different from the speed of light c. Due to this erroneous prediction, the detection of gravitational waves has ruled out several of such modified gravity theories since 2016.

A recent proposal of *varying dark energy* has resulted from a quasar survey published in 2018. It turns out that data from the X-ray space observatories, Chandra from NASA and XMM-Newton from ESA, suggest that dark energy may have varied over cosmic time. The researchers tracked the effects of dark energy out to about 13,000 million years ago (approximately 700 million years after the Big Bang) by determining the distances to 1598 quasars (see Fig. 3.22). These quasars had become standard candles through a new technique that measures their ultraviolet and X-ray radiations; to be precise, this technique measures the relative amounts of these radiations depending on the quasars intrinsic luminosity. The researchers studied the expansion rate

---

[14]Scalar fields, like the Higgs field, have a value in each point of space, but not direction or orientation.

Fig. 3.22 Artist's illustration of a quasar—a quasi-stellar object—which is an extremely bright active galactic nucleus (AGN) powered by a supermassive black hole. Its mass varies between millions and thousands of millions of solar masses, and it is surrounded by a gaseous accretion disk. Quasars release enormous amounts of particles of ordinary matter, antimatter, and radiation, especially through two jets emitted in opposite directions, and represent early stages of some galaxies. *Credit* Robin Dienel, courtesy of the Carnegie Institution for Science

of the Universe back to those very early times and found evidence that the amount of dark energy is growing with time. They also showed that their results agree with the results from supernova measurements over the last 9000 million years, which is the maximum backward time achievable with Type Ia supernovae.

In the models where dark energy is not considered a substance, the acceleration of the Universe's expansion can be achieved by making use of modified gravity theories. But it can also be achieved in the framework of brane worlds, which involve extra dimensions and parallel universes, as explained before (Fig. 3.17). Again, similarly as for dark matter, the influence of the parallel universes in our vicinity could play the role of the "dark energy" responsible for the acceleration of our Universe. In this respect, the *Cardassian Universe*, proposed in 2002 by Katherine Freese (Fig. 3.23) and her Ph.D. student Matthew Lewis, then at the University of Michigan (USA), was one of the earliest of such models. Its only components are matter and radiation, without vacuum energy ($\Lambda = 0$). In spite of this, the Cardassian Universe is flat and experiences an accelerating expansion as a consequence of being a three-dimensional brane in a higher dimensional Universe.

Fig. 3.23 Katherine Freese is a theoretical cosmologist and astrophysicist, and one of the world's leading researchers into dark matter. She has put forward many ideas about its nature—including the existence of "dark stars"—as well as about the early Universe. She is currently Professor of Physics at the University of Texas at Austin (USA), and also Professor of Physics at Stockholm University (Sweden). She has also been Director of the Nordita Institute for Theoretical Physics in Stockholm. In 2019 she received the Lilienfeld Prize, given by the American Physical Society, and in April 2020 was elected to the US National Academy of Science. She has published the outreach book "The Cosmic Cocktail: Three Parts Dark Matter". *Credit* Evan Cohen. Courtesy of Katherine Freese

Finally, there are proposals in which the acceleration of the Universe is not considered a real phenomenon. For example, proponents of the *Inhomogeneous Cosmology* state that the Milky Way is located in an emptier-than-average region of space, so that the observed expansion rate might be misinterpreted as an acceleration. Other proposals claim that the methods employed by the supernova research teams in order to interpret the light-curves were flawed or that they did not have enough statistics. However, according to most experts, a closer look at these proposals reveal that they all have inconsistencies.

3.5.3   Dark Energy and the Destiny of the Universe

The present dominance of the dark energy in the Universe raises the question of its ultimate destiny. This is so because, depending on its nature, the accelerated expansion will or will not continue forever. In the event that dark energy is the cosmological constant Λ, nothing will be able to stop the acceleration; the expansion of the Universe will continue forever, speeding up, faster and faster. The reason for this, as explained above, is that the energy density of Λ does not dilute with the expansion of the Universe, remaining the same, unlike that of matter and radiation, which decrease with the scale factor as a^{-3} and a^{-4}, respectively. As a result, if dark energy were the cosmological constant Λ, once it became dominant over matter and radiation giving rise to the accelerated expansion, this expansion would only help to make it even more dominant, with no turning point.

This leads to a rather dim and gloomy scenario. For one thing, the gas clouds that give rise to the stars would be diluted until no more stars would form. Even though many of them might exist longer than the current age of the Universe, in the end all stars would fade away, and life would be extinguished with them as well. But this would still take millions of millions (10^{12}) of years to occur. In the meantime, the intelligent observers within the galaxies would witness that more and more known objects of their respective skies would disappear from view, as these objects would move farther and farther away turning so dim and redshifted that it would become virtually impossible to observe them. But there is more, because due to the accelerated expansion of space all those objects would really disappear behind each other's cosmological horizons.[15]

This situation would reach the point that, some 500,000 million years from now, each technological civilization would only have access to detect, and hence to investigate, a small set of galaxies that do not drift apart from one another because they are gravitationally bound, being part of the same group—just like Andromeda and the Milky Way are part of the so-called Local Group.[16] Consequently, it would be essentially impossible for those civilizations to discover the expansion of the Universe and therefore its

---

[15]The cosmological horizon, also called particle horizon, defines the size of the observable Universe since it is like a sphere around us whose radius equals the distance to the most distant objects we can observe.

[16]The Local Group contains at least 80 galaxies, most of them dwarf. It consists mainly of two clusters: the Milky Way with its cohort of satellite galaxies and Andromeda with its own. They are separated by a distance of about 0.8 Mpc and move towards each other at a speed of 110 km/s. The group itself is a part of the larger Virgo Supercluster.

beginning in the form of a Big Bang. So, those would be difficult times for Astrophysics, and even worse times for Cosmology.

But a different destiny for the Universe is also possible, depending on the nature of the dark energy. For if it is variable in space and time, and could even flip its sign—turning repulsive forces into attractive ones—then many alternative scenarios are plausible. For example, the attractive gravitational forces could stop the Universe's expansion altogether and reverse the process by triggering a period of contraction. The question now arises as to whether a contracting phase of the Universe would be a better epoch for Astrophysics and Cosmology.... We leave the answer for the interested reader.

4

Discovering Antimatter

During the nineteenth century many experiments with electricity were performed, and not only for scientific reasons but also for entertainment, in salons of the upper social classes as well as on a variety of stages in the world of spectacle. In 1874 George Stoney proposed that electricity came in elementary units that could not be separated from matter, and in 1891, after calculating its magnitude, he proposed the term *electron* to name this unit of electricity. However, a few years later, in 1897 Joseph John Thomson discovered that these units of electricity, the electrons, could be separated from matter; that is, they were corpuscles. This finding was the result of a series of experiments that Thomson carried out at the Cavendish Laboratory of Cambridge University (UK), using cathode ray tubes. Those experiments showed that such rays were nothing but corpuscles with negative electrical charge a thousand times lighter than hydrogen atoms, which were themselves the lightest corpuscles known at that time.

These results represented a milestone of the greatest relevance in the world of Science because of their implications, both scientific and technological. For the first time a subatomic particle was discovered, for the first time an elementary particle was discovered and, moreover, with this it was demonstrated that atoms are not indivisible as some of their constituents can be stripped away from them.

It still took 35 years to identify the first antimatter particle, which turned out to be the antiparticle of the electron, named positron. This was, in addition, the second elementary particle that was discovered besides the photon, since the only subatomic particles known at that time were the electron and the proton, and the latter is not elementary. The discovery of the positron,

© Springer Nature Switzerland AG 2021
B. Gato-Rivera, *Antimatter*,
https://doi.org/10.1007/978-3-030-67791-6_4

announced in 1932 by Carl Anderson, working as a postdoctoral researcher at the California Institute of Technology (Caltech), happened in fact in the late summer of 1931, one year after he had finished his doctoral studies (he received the Ph. D. degree in June 1930). Amazingly, the existence of the electron's antiparticle had just been proposed by the physicist Paul Dirac, at Cambridge University, in an article sent to the journal "Proc. of the Royal Society A" in March 1931, published September 1. In that article Dirac tried to make complete sense of the equation he had written down in 1927, published in 1928, in order to describe electrons moving at very high speeds. But neither Anderson was following Dirac's work nor the other way around, even though Anderson knew of the existence of Dirac's theory, which he, like many other physicists at that time, regarded as rather incomprehensible, and "highly esoteric". Yet, within a few weeks or days, Anderson was taking pictures of the traces left by the particle whose existence was postulated in Dirac's article.

But let us leave this story for now and resume it when we return to the discovery of the positron after introducing the Dirac equation. Then, we will briefly review the findings of the antiproton and antineutron and the production of antinuclei and antiatoms, finishing with a section about primordial versus secondary antimatter.

4.1   The Dirac Equation

As we said, the existence of the positively charged "electron" had been postulated in 1931 by Paul Dirac (Fig. 2.7 and Fig. 4.1). This prediction was a consequence of the equation he had proposed in 1927 to describe electrons in the relativistic regime—at high speeds close to the speed limit c. This equation can be expressed as:

$$i \, \gamma^{\mu} \, \partial_{\mu} \, \psi = m \, \psi, \tag{4.1}$$

where ψ is the quantum field of the electron and m its mass. Now, despite its apparent simplicity, this equation is quite complicated because ψ has four components, and the so-called Dirac gamma matrices γ^{μ}, which are four, have 16 components each.

One of the merits of this equation is that it also predicts that the electron has an intrinsic angular momentum, which is called spin, as we saw in Sect. 2.4.1. This was very important because it could explain the doubling of the available states for the electrons at the different energy levels of the atoms. Yet spin was not the only novelty this equation predicted, as it turned

Fig. 4.1 Paul Dirac (1902–1984) at work at Cambridge University around 1930.
Credit AIP Emilio Segrè Visual Archives, Gift of Mrs. Zemansky

out that the Dirac equation, besides describing electrons, also has solutions corresponding to negative energies. These solutions baffled both Dirac and his close colleagues because they did not know how to interpret them.

Then, on a first attempt, he published the article *A Theory of electrons and protons*, in the "Proceedings of the Royal Society A", which appeared on January 1, 1930. In this article he put forward his "Sea" theory (see below) and claimed that "*an electron with negative energy moves in an external electromagnetic field as though it carries a positive charge*". Dirac further argued that the mass associated to the negative energy electron was not necessarily the same as that of the ordinary, positive energy, electron. Accordingly, *he proposed the proton* as the candidate for that positively charged "electron", even though experiments showed that protons were 1836 times heavier than electrons and his equation did not offer a good reason for such a large discrepancy.[1] In any case, to fully understand Dirac's thinking one has to take into account that the only subatomic particles known at that time were electrons and protons; neutrons were discovered shortly afterwards, in 1932, although

---

[1] He stated, with respect to the positive and negative energy solutions: "The symmetry is not, however, mathematically perfect when one takes interaction between the electrons into account….The consequences of this dissymmetry……we may hope it will lead eventually to an explanation of the different masses of proton and electron".

their existence had been already postulated by Rutherford in 1920 in order to explain the stability of most atomic nuclei.

Immediately after the publication of Dirac's article, Robert Oppenheimer (Fig. 4.2), then at Caltech, reacted very strongly against that proposal sending a note to the journal Phys. Rev. Lett., received on 14 February 1930. He argued that if protons corresponded to the negative energy solution of the Dirac equation, i.e. to positively charged "electrons", then the hydrogen atom, and all kinds of atoms, would rapidly destroy themselves because the positively and negatively charged electrons would meet and would annihilate with each other. He made some calculations that resulted in "*a mean lifetime for*

Fig. 4.2 Robert Oppenheimer (1904—1967), was gifted with a deep insight and great intuition that allowed him to make many outstanding contributions to several branches of Physics such as Quantum Mechanics, Nuclear Physics, Particle Physics, Cosmic Rays and Astrophysics. With two of his students he investigated the gravitational collapse of neutron stars and black holes many years before they were discovered. He became the leader of the Manhattan atomic bomb project in Los Alamos Laboratory despite his past left-wing activities and his close association to members of the Communist Party. © Copyright Triad National Security, LLC

ordinary matter of the order of 10^{-10} *seconds*". But he also asserted that "*If we return to the assumption of two independent elementary particles, of opposite charge and dissimilar mass, we can resolve all the difficulties raised in this note......*". Hence, Oppenheimer did not predict the existence of the anti-electron, whose mass must be exactly the same as that of the electron; his point was rather that the positively charged electron of Dirac could not be the proton, but it could be another yet unknown particle whose mass need not be the same as the electron mass.

Dirac understood and accepted Oppenheimer's criticism. In addition, he realized, with the assistance of Hermann Weyl and his brand new book "Group Theory and Quantum Mechanics" (in German, 1931), that the mass of the positive and negative energy solutions to his Eq. (4.1) had to be the same. Then, he published yet another article in the "Proceedings of the Royal Society A", in 1931, received on 29 March and released on September 1, where he stated, in relation with the negative energy states:

> The question then arises as to the physical interpretation of the negative-energy states, which on this view really exist.... Subsequent investigations, however, have shown that this particle necessarily has the same mass as an electron and also that, if it collides with an electron, the two will have a chance of annihilating one another much too great to be consistent with the known stability of matter...... would be a new kind of particle, unknown to experimental physics, having the same mass and opposite charge to an electron. We may call such a particle an anti-electron. We should not expect to find any of them in nature, on account of their rapid rate of recombination with electrons, but if they could be produced experimentally in high vacuum, they would be quite stable and amenable to observation. An encounter between two hard γ rays (of energy at least half a million electron volts) could lead to the creation simultaneously of an electron and an anti-electron, the probability of occurrence.....

Therefore, Dirac decided to postulate that the negative energy states of his equation corresponded to a particle identical to the electron, but with the opposite electrical charge, particle that he named *anti-electron* because it would be annihilated upon contact with an electron. Moreover, he deduced that pairs of electrons and anti-electrons could be created from the energy of pure γ rays, provided that this energy reaches the threshold corresponding to two electron masses (511 keV/c^2 each), as given by the Einstein formula of mass-energy conversion $E = m\,c^2$ (see Fig. 2.2 in Chap. 2).

Furthermore, in the same article Dirac suggested that: "*Presumably the protons will have their own negative-energy states, ...one appearing as an anti-proton*". Clearly, with that statement the concept of *antiparticle* was born and, by extension, also the concept of *antimatter*. In Fig. 4.3 one can see a commemorative plaque honoring Paul Dirac and his equation, one of the most important equations in Physics ever, placed in 1995 at Westminster Abbey in London (England), adjacent to Newton's grave.

Curiously, the terms antimatter and antiatoms had been coined more than 30 years before, in 1898, by the physicist Arthur Schuster, but with the meaning of "gravitationally repulsive" matter. Notice that only one year earlier the electron had been discovered, and the internal structure of atoms was still unknown, let alone that many scientists did not even believe in the existence of atoms. Schuster published the letter *Potential Matter—A Holiday Dream*, in the journal "Nature", but it was presented more as a science fiction story than as a scientific article. In this letter he reflected on the possibility of repulsive gravitation. He was wondering:

> ...another set of atoms, possibly equal to our own in all respects but one; they would mutually gravitate towards each other, but be repelled from the matter which we deal with on this earth". "Worlds may have formed of this stuff, with elements and compounds possessing identical properties of our own....; different only in so far that if brought down to us they would rise up into space with an acceleration of ....if they ever existed on our earth, they would long have been repelled by it and expelled from it". "The atom and the anti-atom may enter into chemical combination, because at small distances molecular forces would overpower gravitational repulsions.... Matter and anti-matter may

Fig. 4.3 Commemorative plaque in honor of Paul Dirac and his equation at Westminster Abbey in London. It was placed on the floor, adjacent to Newton's grave. O.M means "Order of Merit", recognizing distinguished service in the armed forces, science, art, literature, or for the promotion of culture. The plaque was placed in 1995, eleven years after Dirac's death. *Credit* Courtesy of CERN

further coexist in bodies of small mass.....When the atom and anti-atom unite, is it gravity only that is neutralized, or inertia also? May there not be, in fact, potential matter as well as potential energy?

At this point it should be noted that Dirac's interpretation of the nature of the antiparticles, put forward in the aforementioned 1930 article, is called the *Dirac Sea* theory and is used nowadays in some branches of Physics, like Solid State Physics. According to this theory, the quantum vacuum would be like an infinite sea filled with negative energy particles, and the antiparticles would be like holes in this sea, like vacancies left by these particles. These holes would behave exactly like ordinary particles—for example, under the action of an electric field—but with the opposite electric charge. The Dirac Sea theory was superseded by Quantum Field Theory, where particles and their antiparticles are viewed at the same footing, i.e. as perturbations of the same quantum field, and the vacuum is simply the absence of any particles. Indeed, we and our entire world could be made up of antimatter instead of matter, as will be discussed in Chap. 7 and in the Epilogue.

4.2   The Positron

Carl Anderson (Fig. 4.4) had been investigating X-rays for several years at Caltech, under the supervision of the prestigious Nobel Prize laureate Robert Millikan. For his research he was using a cloud chamber, also called Wilson chamber after Charles Thomson Wilson, who invented it (see Fig. 4.5). Then, in 1930, while still a graduated student, Anderson decided to start a new experiment using again a cloud chamber, but this time operated in a magnetic field. The experiment consisted in studying the interaction of γ rays, much more energetic than X-rays, with matter. To be precise, he wanted to study the energy distribution of the electrons produced by the impact of the γ rays on a thin lead plate inserted inside the chamber. The source of the γ rays was the radioactive isotope of thallium ^{208}Tl, misclassified at that time as a thorium isotope denoted as ThC'. So, this substance had to be placed inside the cloud chamber as well, and the idea was to take photographic plates of the traces that the electrons thus produced would leave as they moved throughout the chamber.

Fig. 4.4 Portrait of Carl David Anderson (1905–1991) published in "The Science News-Letter", in December 1931, along with the first photograph of the track of a positron (Fig. 4.7). This portrait was also used by the Nobel Foundation when he received the 1936 Nobel Prize in Physics (shared with Victor Hess), given for the positron discovery by studying the cosmic radiation with a Wilson cloud chamber in a very strong magnetic field. This was the first antimatter particle to be identified. The finding was officially confirmed in August 1932, after passing the "lead plate" test, and the name "positron" was put forward by Anderson six months later in a very detailed article entitled *The Positive Electron*. In 1936 he discovered the muons, along with Seth Neddermeyer, although they had photographs with muon tracks since 1931. *Credit* Bortzells Esselte, courtesy AIP Emilio Segrè Visual Archives, Weber Collection, W. F. Meggers Gallery of Nobel Laureates Collection

Wilson Cloud Chambers and Bubble Chambers

The Wilson cloud chambers are filled with a supercooled gas that gets ionized—gaining or losing electrons—when interacting with a charged particle passing by, and once ionized it condenses leaving a visible trace of the particle's

Fig. 4.5 Charles Thomson Wilson (1869–1959) when he received the 1927 Nobel Prize in Physics for his invention of the cloud chamber, in 1911, which made possible the study of subatomic particles, although it was not until 1923 that this device was developed sufficiently to be used for scientific research. He was the first person to observe and photograph traces of alpha particles and electrons, and probably also traces of positrons, since they are undistinguishable from those of electrons in the absence of a magnetic field. Despite several important contributions to Particle Physics, Wilson remained all his career particularly interested in atmospheric physics, especially atmospheric electricity and cloud formation. On the right, one can see a Wilson cloud chamber at Brookhaven National Laboratory (USA), that was used for cosmic ray research around 1955. *Credit* The Nobel Foundation (left) and Wikimedia Commons (right)

trajectory. In addition, cloud chambers are usually placed in magnetic fields that make charged particles rotate producing characteristic spiral trails. In this way it is easy to distinguish positively charged particles from negatively charged ones, since they turn in opposite directions. Moreover, the curvature of the particle's trajectory depends not only on the strength of the magnetic field but also on the particle's speed and mass. On one hand, heavier particles require more energy to be bent than lighter ones and, on the other hand, for a given species of particle, slower ones move in tighter curves than faster ones since they are easier to bend by the magnetic field.

Another piece of information that allows to identify charged particles crossing through the chamber is the ionization density of the traces, which depend on the quotient between the mass and the energy of the particles. So, for a given particle the traces are thicker when they move slower and for particles with the same energy but different masses the traces are thicker for the more massive ones. Neutral particles—mainly photons and neutrinos—lacking electric charge leave no trace in the chamber, but it is easy to infer their existence from the traces left by the charged particles using conservation principles of some properties, like total energy and momentum, which apply in all processes between particles (see Fig. 2.5 in Chap .2).

Cloud chambers made the study of subatomic particles possible and played an essential role in experimental Particle Physics from 1923 until the advent of the bubble chambers in the 1950s. The working principle of the bubble chamber, invented by Donald Glaser in 1952, is very similar to that of the cloud chamber, but it makes use of a superheated liquid—usually liquid hydrogen—that vaporizes producing tiny bubbles around the ionization traces left by charged particles, making them visible. Bubble chambers, that allow much larger sizes than cloud chambers, were the most used particle detectors for many decades, after cloud chambers were effectively superseded at the beginning of the 1960s. They have a very limited used nowadays, however, because they have been superseded, in turn, by silicon detectors, wire chambers, spark chambers and drift chambers.

Now, before the definitive cloud chamber was built for that experiment, Anderson carried out some tests with his own chamber without putting it in a magnetic field. The first electron traces he obtained were very faint and difficult to photograph due to the higher speeds of the electrons produced by the impact of the γ rays, as compared with the much slower electrons produced using X-rays. After several attempts to make the traces more visible, he got the idea of pouring some ethyl alcohol in the water of the chamber and luckily this solved the problem. Then he was ready to properly design and build an upgraded cloud chamber specific for the γ rays emitted by the ThC' isotope.

However, before he started, Anderson received a very special and apparently urgent request by his supervisor. This was to build a cloud chamber designed specifically to measure the energies of the electrons present in the cosmic radiation, i.e. the secondary cosmic-ray electrons produced in the atmosphere by the impact of the primary cosmic rays coming from outer space. It turned out that a few years earlier, Dmitri Skobeltzyn,[2] working in the Leningrad Polytechnical Institute (USSR), had discovered straight tracks of very energetic cosmic-ray particles in photographic plates from a Wilson

---

[2]Dmitri Skobeltzyn was in 1923 the first scientist to observe, in a Wilson cloud chamber, the Compton effect—the interaction of the γ rays emitted by a radioactive substance with the electrons of matter. In 1924, he was also the first to operate the cloud chamber in a magnetic field (from 1000 to 2000 gauss). In addition, he was the first physicist to observe, since 1925, straight tracks left by very energetic cosmic-ray particles while passing through a cloud chamber, and the first to demonstrate that indeed cosmic-ray particles can be highly energetic. He published the first image with the track of such an energetic particle in 1927 and a full report of his results in 1929. More details in Chap. 5 and Appendix B.

cloud chamber operated in a magnetic field, and this line of research was gaining increasing interest (see Fig. 5.2 in Chap. 5 and Figs. B.1–B.3 in Appendix B).

In addition, Millikan was an expert in cosmic rays and firmly believed in his own hypothesis about their nature and about the mechanism for the production of the secondary electrons, which did not permit energies above 500 MeV (million electron volts). But he was also a very busy man who needed collaborators to be able to engage in any scientific research, and Anderson had shown his excellent skills to design and build cloud chambers. So, Millikan wanted him to investigate the energy spectrum of the cosmic-ray electrons in order to test his hypothesis. To this end, he promised Anderson a National Research Council fellowship—the only fellowship available at that time for postdoctoral studies—provided he accepted to devote himself exclusively to Millikan's research project on cosmic rays.

This new project included providing the cloud chamber with the strongest magnetic field available in order to bend the tracks of the cosmic-ray electrons, which were expected to be "very" energetic—several hundred MeV. These energies were much higher than the energies of the electrons—at most 1 MeV—previously obtained working with γ rays emitted by radioactive substances. Thus, the cosmic-ray electrons were much harder to deflect and the cloud chamber had to be placed inside a very strong magnetic field. With the help of engineers at the nearby Guggenheim Aeronautical Laboratory, on the Caltech campus at that time, Anderson could build that powerful magnet after several months of intensive work. It could reach up to 24,000 gauss working at full power, which is about 100,000 times stronger than the Earth's magnetic field at the equator. In the summer of 1931, the cloud chamber— 17 × 17 × 6 cm in size—began operation inside the bulky magnet, with its long dimension vertical, and the whole apparatus was dubbed "magnet cloud chamber" (Fig. 4.6).

Using this magnet cloud chamber, Anderson realized from the very beginning that some tracks, which apparently corresponded to electrons, were bending the other way, indicating positive electric charge. At first, he thought that those particles were probably protons, but soon it became clear to him that the ionization they produced, as well as the curvature of their trajectories, corresponded to particles with masses very similar to that of electrons but with positive electric charge. But let us Anderson himself report about this episode[3]:

---

[3]The following extracts are taken from Anderson's contribution *Unraveling the Particle Content of the Cosmic Rays*, and Skobeltzyn's contribution *The Early Stage of Cosmic Ray Particle Research*, to the book "The Birth of Particle Physics", from 1983, based on the lectures and round-table discussions

Fig. 4.6 Carl Anderson with the magnet cloud chamber with which he discovered the positron e^+ as well as the muon μ^- and antimuon μ^+. This picture was taken in the Guggenheim Aeronautics Laboratory in 1931. Anderson operated this apparatus also in high-altitude cosmic ray experiments, first carried out in the mountains, on the summit of Pike's Peak in 1935, and later on board a B-29 airplane, in 1946. *Credit* Courtesy of Caltech Archives

> The first results from the magnet cloud chamber were dramatic and completely unexpected. There were approximately equal numbers of particles of positive and negative charges, in sharp contrast to the Compton electrons expected from simply the absorption of high-energy photons. Dr. Millikan was on a visit to England at the time the first results were obtained, and I sent him a group of 11 photographs.... Only a few of the low-velocity particles were clearly identified as protons.

Indeed, in the autumn 1931 Millikan visited Europe and gave lectures in Paris and Cambridge, showing the 11 photographs he had received from Anderson. However, he ascribed the positive "anomalous" tracks to high-energy protons produced by high-energy cosmic-ray photons; to be precise, these energetic protons would result from the interaction of ultra energetic γ rays with atomic nuclei. This explanation was in line with Millikan's hypothesis about the nature of the cosmic rays (γ rays produced by some atom-building processes in outer space). Moreover, these ideas seemed to be very firmly fixed in his mind, according to Anderson, who actually did not

---

at the "International Symposium on the History of Particle Physics", held at the Fermi National Laboratory (Fermilab, near Chicago, USA) in May 1980.

support Millikan's hypothesis but was forced to do so in articles co-authored by both of them.

Now, it turned out that Skobeltzyn, who could not attend Millikan's talks, got detailed reports about the photographs presented by Millikan in letters by Marie Curie and Frederic Joliot-Curie from Paris and Harold Gray from Cambridge. Let us see his reaction with respect to the anomalous positive tracks found by Anderson:

> After I received Joliot's letter, I wrote him straightforwardly my views on the subject and suggested that something was wrong with Millikan's photos or their interpretation.... Now, the ionization density (specific ionization) depends mainly on the velocity of the particle..... It follows that the ionization produced by a proton of 50 MeV is practically the same as that of an electron of about 25 keV. It is impossible to confuse the specific ionization of such a slow electron with the ionization of a fast ("relativistic") one, having several MeV or more. However, the positive tracks of pictures demonstrated by Millikan did not differ essentially from electron tracks on the same pictures, with the energies of about 50 MeV. Prof. Millikan and his audience overlooked this inconsistency.... In their article of May 1932, Millikan and Anderson repeated the interpretation Millikan had made in Europe.... Progress by Anderson in the deciphering of his experimental evidence was slow. Only in September 1932 (a year after a stock of more than 1000 Wilson photos in a strong magnetic field had been obtained) did he make reference, in a brief and very cautiously drafted note, to the specific ionization of the positively curved tracks, and he concluded the existence of positively charged particles, the mass of which must be small compared with the mass of a proton.

Thus, Skobeltzyn shared exactly the same views of Anderson that the positive tracks looked very much like those of electrons—except for the opposite electric charge—and could not possibly be ascribed to protons (see Fig. 4.7). He also reflected on *"the psychological barriers lying in the way of discovering the first of a sequence of many generations of new particles...."* Not only that but, at that time, only electrons, protons and photons were known, and no other particles were necessary to explain the world (except perhaps for the neutron, which was in fact discovered in 1932).

In any case, despite the pressure exerted by Millikan, Anderson somehow managed to start advertising his "positive electron" tracks as early as December 1931, when a photo entitled *Cosmic Rays Disrupt Atomic Hearts* appeared in "The Science News-Letter" (Fig. 4.7, left). The caption explained some of the results obtained by Anderson and the photo showed two similar

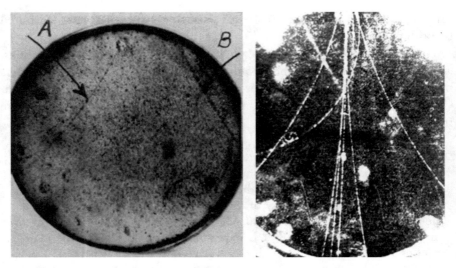

Fig. 4.7 Photographs of positron and electron tracks from the cosmic radiation obtained by Carl Anderson in 1931-1932 using his magnet cloud chamber. On the left, one can see the first ever published photograph where a positron track could be identified (B), together with an electron track (A). It appeared in "The Science News-Letter", on 19 December 1931. On the right, a group of particles coming down; electrons bending to the left and positrons to the right. One can distinguish at least two electron–positron pairs where the electrons and positrons leave very similar traces which are mirror images of each other. Most tracks in this and other photographs seem to have a common origin, so Anderson rediscovered the cosmic ray showers, after Skobeltzyn, and concluded that such tracks are from secondary particles originating in collisions, not from cosmic-ray particles coming directly from outer space. *Credit* Courtesy of Caltech Archives

traces but curving differently, one from an electron and the other from a positive charged particle, described by the science journalist as "*probably a proton*" (but not *surely a proton*, as Anderson explained to him).

About that time, Neddermeyer joined Anderson as his first graduate student, and he was assigned the task of continuing the curvature measurements of the traces, paying particular attention to obtaining as precise measurements as possible for those of highest energy, in the range above 1 GeV (1000 MeV), which completely crushed Millikan's hypothesis, although he was very difficult to convince:

> One of the first tasks undertaken with the first photographs, in fact the original purpose of the experiment, was to determine an energy distribution of the particles by means of the curvature they showed in traversing the powerful magnetic field. My original measurements showed an energy distribution extending from very low energies of about 100 MeV up to above 1 GeV,

with the great majority of particles having energies in the range of several hundred MeV. ... On many occasions Neddermeyer and I would meet with Millikan to discuss energy measurements and their interpretation. According to his hypothesis, the energies of the electrons observed should be in the general energy range of about a hundred million electron volts, but not to exceed some 400 to 500 MeV. I remember that on one occasion Neddermeyer was relating energy measurements he had made on a series of tracks and he came to one over 1 GeV. Millikan virtually hit the ceiling and gave Neddermeyer a rather tough third-degree-type questioning. Both Neddermeyer and I tried to argue with Millikan, but it seemed impossible to change the direction of his thinking—his mind's momentum seemed close to infinite. It was only after many of these meetings that Millikan readily accepted energies in the range of several GeV. (Note added: at that time, the GeV (10^9 eV) was denoted as BeV, here we have updated the notation.)

From the previous paragraphs, it transpires that Anderson had a hard time trying to convince Millikan, his research supervisor and collaborator, that the anomalous positively charged traces were not made by protons crossing the cloud chamber, but by much lighter particles similar to electrons. Fortunately, Anderson had the courage to publish this finding alone; but before going public, he had to make sure that the traces he had found did not correspond to electrons moving upwards, instead of downwards. It happens that cosmic-ray particles from a cascade often hit molecules near the ground, and some of the resulting radiation is bounced upwards. If this had been the case, the electron tracks coming from below would indeed look like those of positively charged particles coming from above. As a matter of fact, that was precisely the first explanation that had come to Anderson's mind until he realized that there were too many electrons bouncing upwards.

To this end, Anderson placed a 6 mm lead plate in the center of the chamber in order to reduce the speed of the particles passing through it, resulting in a tighter curvature of their trajectories that could be seen with the naked eye. Shortly after, the first "positive electron" made its appearance, as shown in Fig. 4.8, leaving a clear track of its passing through the lead plate (the central horizontal structure). Curiously, it turned out that this first specimen actually came from below, moving upwards, as one can easily deduce by inspecting the change in the curvature of the track, which is bigger on the upper part. This photograph was taken by Anderson on August 2, 1932, and was released first in a brief report in "Science", in September 1932. Afterwards it was published in the article *The Positive Electron* in "Physical Review", on 15 March 1933, where he presented his discovery in depth. In the abstract, the editor already stated: "*These particles will be called positrons*",

Fig. 4.8 On the left is the photograph obtained by Carl Anderson with his magnet cloud chamber on August 2, 1932. It shows the trail of the first confirmed positron having passed the "test" of the lead plate. It came from below upwards as can be deduced from the curvatures before and after crossing the plate. Namely, its energy before it traverses the plate is 63 MeV and after it emerges its energy is 23 MeV. The length of the path after crossing the lead plate is at least ten times greater than the possible length of a proton path of this curvature. On the right is the Wilson cloud chamber used by Carl Anderson and Seth Neddermeyer between 1935 and 1941 at Caltech. The chamber has a diameter of 15.87 cm. *Credit* Physical Review 43, 491, 1933 (left) and Historical Instrumentation Collection of the Health Physics Museum, Oak Ridge (USA) (right)

as had been suggested by Anderson in the text: "*…the positive electron which we shall henceforth contract to positron…*".

Among other things, in that article Anderson fully explained that the length of the upper path, corresponding to the particle after traversing the lead plate, was at least ten times greater than the possible length of a proton path of this curvature, thereby discarding that possibility once and for all (Fig. 4.8). Other relevant excerpts from that article were:

> In the course of the next few weeks other photographs were obtained which could be interpreted logically only on the positive-electron basis…. It is concluded, therefore, that the magnitude of the charge of the positive electron which we shall henceforth contract to positron is very probably equal to that of a free negative electron…. To date, out of a group of 1300 photographs of cosmic-ray tracks 15 of these show positive particles penetrating the lead, none of which can be ascribed to particles with a mass as large as that of a proton, thus establishing the existence of positive particles of unit charge and of mass small compared to that of a proton….. From the fact that positrons occur in groups associated with other tracks it is concluded that they must be secondary particles ….

Furthermore, Anderson mentioned that recent press reports had announced that Patrick Blackett and Giuseppe Occhialini (Fig. 4.9), at the Cavendish Laboratory in Cambridge (UK), had also obtained evidence for the existence of light positive particles in an extensive study of cosmic ray tracks, confirming his previous results published in Science. Not only that, but Blackett and Occhialini, using a cloud chamber larger and more sophisticated than that of Anderson, had also discovered the *inverted V* shaped tracks and the *spiral* tracks revolving in opposite directions, all of them left by the production of electron–positron pairs from γ rays. In Fig. 4.10 one can see a beautiful photographic plate of such tracks obtained many years later at the Lawrence Berkeley National Laboratory (USA) using a bubble chamber.

Fig. 4.9 Patrick Blackett (1897–1974), on the right, and Giuseppe Occhialini (1907–1993), around 1933. They worked in the group of Rutherford at the Cavendish Laboratory in Cambridge (UK) and made an extensive study of cosmic ray tracks, confirming the existence of the positron. They discovered the "inverted V" shaped tracks and the spiral tracks revolving in opposite directions, left by the production of electron–positron pairs from γ rays. They also rediscovered the cosmic ray showers, after Skobeltzyn and Anderson. *Credit* Giuseppe Occhialini and Constance Dilworth Archive, Università degli Studi di Milano

Fig. 4.10 On the left is a photographic plate made in a bubble chamber at the Lawrence Berkeley National Laboratory (USA) showing two processes of electron–positron pair creation. A very energetic photon penetrates from the top (being invisible as it lacks electric charge), collides with an atom's electron and pulls it out, as we see in the trace left by its trajectory; but part of the collision's energy is invested in creating a pair $e^+ e^-$, whose trajectories tightly revolve drawing spirals in opposite directions. The photon then creates another more energetic $e^+ e^-$ pair, as can be seen from the "inverted V" shaped pair of tracks due to the smaller curvature of the trajectories. *Credit* Lawrence Berkeley National Laboratory. On the right, the Big European Bubble Chamber (BEBC) at CERN. It started operation in 1973 and was decommissioned in 1984, having delivered 6.3 million photographs to 22 experiments. At present it is on display at the garden of CERN's Microcosm museum. *Credit* Courtesy of Fanny Schertzer and Wikimedia Commons (Creat. Comm. 2.0 license)

That was the definitive proof that the positrons discovered by Anderson were in fact the anti-electrons postulated by Dirac, that is, particles with the same mass and spin as the electrons but with the opposite electric charge. Therefore, the experimental findings of Anderson, Blackett and Occhialini confirmed the existence of Dirac's antimatter in 1933. But these findings were also the proof that photons—pure energy without mass—could be converted into massive particles; in particular into a particle-antiparticle pair according to the formula $E = m c^2$, as Dirac had already predicted in his 1931 article where he put forward the existence of the anti-electron. Thus, this was a triumph for Einstein and his theory of Special Relativity as well. In this respect, in his 1980 reminiscences for the book mentioned in footnote 3, Anderson states:

If one goes back a few years, say to just after the Dirac theory was announced, it is interesting to speculate on what a sagacious person working in this field might have done. Had he been working in any well-equipped laboratory, and had he taken the Dirac theory at face value, he could have discovered the positron in a single afternoon. The reason for this is that the Dirac theory could have provided an excellent guide as to just how to proceed to form positron-electron pairs out of a beam of γ-ray photons.

But there is more because, in early 1933, Anderson resumed with Neddermeyer the experiment he had initiated in 1930 and had to abandon at Millikan's request, as explained before. As a result, they achieved the first direct evidence that the γ rays emitted by the radioactive isotope ^{208}Tl, misclassified at that time and denoted as ThC', can also give rise to electron–positron pairs by interacting with some materials. In fact, they could see that about 10% of the "electrons" emerging in this experiment had a positive charge. So, Anderson concluded that the discovery of the positron could have taken place already in 1930 if he had had the opportunity to perform that experiment. Soon afterwards, in the spring of 1933, the same results were obtained, independently, by the Joliot-Curies and by Lise Meitner and her assistant Kurt Philipp.

Curiously, it turned out that Skobeltzyn conducted a similar experiment with ThC' two years before, in early 1931, and noticed the anomalous traces of particles similar to the electron but bending the other way around under the action of the magnetic field.[4] He was puzzled and had no idea how to interpret those weird tracks, until Anderson came up with the explanation one year later. It must be clarified that the cosmic-ray tracks he had recorded already in 1925 were only straight lines, showing no bending at all, because the magnetic fields he was using (from 1000 to 2000 gauss) were not strong enough to bend the trajectories of such energetic particles. Surprisingly, Skobeltzyn had no interest whatsoever in using much stronger magnetic fields capable of bending such tracks because these fields were unnecessary for his main topic of research; otherwise, he might have discovered the positron in the cosmic radiation around 1926.

Actually, among the collaborators of Skobeltzyn, the only one who showed interest in bending the tracks of the cosmic-ray particles using much stronger magnetic fields was Pierre Auger in early 1931, but he failed miserably, as related by Skobeltzyn:

---

[4]There is a false belief, especially among British physicists, that Skobeltzyn used to see positron traces that puzzled him much before 1930, even though Skobeltzyn himself declared more than once that this was not true, although indeed it happened in 1931. In Appendix B we will have a close look at this myth.

It is well known what extremely important results followed from further development of the technique of using a Wilson chamber plus a magnetic field. The next step that appeared natural was to use a magnetic field of much higher strength. There were many reasons why I myself never tried to do this. In 1929-31, I was working at the Curie laboratory in Paris. Pierre Auger, who had been working at the neighboring institute of Jean Perrin, asked me (probably at the beginning of 1931) if I intended to undertake such investigations. I answered in the negative, whereupon he told me that he would try to perform that kind of experiment. Soon thereafter, he showed me his installation that was ready for operation. However, his attempt turned out to be unsuccessful and probably was dropped by him in the fall of 1931 when it was disclosed that Carl David Anderson had already obtained some thousand beautiful pictures of cosmic-ray tracks in a strong magnetic field (13,000 gauss).... It seems that something was wrong with the cloud chamber that Auger had constructed. I was told in the spring of 1931 that his Wilson chamber, when put into operation, showed no cosmic-ray tracks whatsoever. I obtained this information from a fellow of the Curie laboratory staff (Georges Fournier). ... Until the end of my sojourn in Paris (1931), I had no further occasion to meet Auger himself, and afterward I never heard from him what had gone wrong with his cloud chamber in a strong magnetic field.

Anderson's achievements, especially the discovery of the positron together with some other contributions concerning the nature and properties of the cosmic radiation, earned him the 1936 Nobel Prize in Physics, which he shared with Victor Hess. Curiously, Anderson was only 31 years old at the time, the same age as Werner Heisenberg and Paul Dirac when they received the Nobel Prize in Physics, in 1932 and 1933, respectively, the three of them being the youngest physicists in history to receive such recognition. The discovery of the positron was rightly attributed to Carl Anderson because he was the first scientist to identify this particle and to publish its finding, although several other scientists were very close, especially Skobeltzyn and Patrick Blackett with Giuseppe Occhialini. Let us take a closer look at this.

To start with, it turns out that positrons are also spontaneously emitted by some radioactive substances, independently of the fact that some of the released γ rays may also give rise to electron–positron pairs, as in the case of the isotope ThC'. In addition, since 1930 a number of experiments were carried out in which light atomic nuclei, like those of beryllium, boron, and lithium, were bombarded by alpha particles from polonium or other radioactive sources. As was shown later by Chadwick, Blackett and Occhialini, this simple technique produces neutrons and γ rays, as well as induced radioactivity, but the γ rays also give rise to electron–positron pairs. However, in such experiments no cloud chambers were used, or they were used without a magnetic field, so that the discovery of the positron would have been impossible (at least, until 1932).

Positrons are spontaneously emitted by some radioactive atomic nuclei, and they constitute the so-called β^+ *radiation*, which can be regarded as the "inverse" of the β radiation of electrons, that was explained in detail in Chap. 2 (Sects. 2.4.4 and 2.5.3). This is so because in the case of the β^+ radiation what is transformed inside some unstable nuclei is a proton, due to its interactions with other particles, becoming a neutron and giving as by-products one positron e^+ and one electron neutrino ν_e:

$$p \rightarrow n + e^+ + \nu_e. \tag{4.2}$$

Let us remember that the β radiation, much more frequent, results from the decay of a neutron inside the nucleus, becoming a proton and producing one electron e^- and one electron antineutrino $\overline{\nu}_e$, as was shown in (2.10). In the β^+ radiation, however, the process (4.2) is not really a decay of the proton but a decay of the nucleus. The reason is that a genuine proton decay would require the final particles to be lighter than it, which is not the case here as the neutron is heavier than the proton. This apparent "proton-decay" occurs due to the interactions of the proton with other particles, and it also takes place in nuclear reactions inside stars, as was pointed out in Sect. 2.5.3.

Anyway, as the finding of the positron by Anderson, Blackett and Occhialini spread around, other physicists and chemists took a careful look at their cloud chamber photos searching for traces of positrons. As expected, in addition to Skobeltzyn, several other scientists had also recorded positron tracks in 1931 or 1932—but not earlier—which they had overlooked or mistaken for tracks of protons. Thus, those scientists missed the discovery of the positron for not paying enough attention to the traces recorded on their cloud chambers. Among those scientists were Marie Curie, Lise Meitner, and the "Joliot-Curies" (Irène Joliot-Curie, daughter of Marie Curie, and her husband Frédéric), who had just missed the discovery of the neutron.[5] Nevertheless, they took revenge because only two years later, in 1934, they succeeded in creating artificial radioactive isotopes of phosphorus, which do not exist in nature and also emit positrons. This was achieved by bombarding

---

[5]In January 1932, the Joliot-Curies published in the French journal "Comptes Rendues" the first results of their work on beryllium bombarded by α particles. They reported that only gamma radiation was released from the beryllium, with a particularly powerful energy which was transferred to the protons of a block of wax. James Chadwick, at the Cavendish Laboratory, and Ettore Majorana, in Rome, immediately realized that the authors had mistaken an unknown neutral particle for gamma radiation. Then Chadwick performed in just a few weeks a series of experiments showing that the γ ray hypothesis was untenable. He sent a letter to Nature, received on February 17, 1932, with the title *Possible existence of a neutron*, and he received the 1935 Nobel Prize in Physics for this discovery.

stable aluminum nuclei with α particles, and they were so successful that they continued creating other radioactive isotopes by bombarding other stable nuclei. They called this phenomenon *artificial radioactivity* and for their discovery they were awarded the Nobel Prize in Chemistry in 1935.

Also to be mentioned is the case of the young physicist Chung-Yao Chao, another graduate student at Caltech during the period 1927–30, like Carl Anderson, working under the supervision of Robert Millikan as well, in a room close to that of Anderson. He was studying the absorption and scattering of γ rays, emitted from the radioactive isotope ThC' (i.e. ^{208}Tl), using an electroscope, and he noticed that both the absorption and scattering of the γ rays in lead were substantially greater than calculated by the "canonical" Klein-Nishina formula. A detailed explanation of these anomalous effects was not possible from his experiments because one could not obtain more detailed information from an electroscope, which is a rather rudimentary instrument that only gives a rough indication of the quantity of electric charge.

However, Chao did not have the initiative to repeat his experiments using a cloud chamber (neither in Caltech nor later in China), even though he knew very well the investigations carried out by Anderson using such a device. It is also very likely that Anderson would have helped Chao to build his own cloud chamber if he had asked Anderson to do so. Furthermore, Chao did not make any conjecture or proposal about the possible existence of new particles in order to explain his observations. By contrast, it was Anderson who had the initiative to repeat the experiments of Chao using a cloud chamber, once he was relieved from his X-ray studies for his thesis. But he could perform those experiments only in 1933, with Neddermeyer, finding a remarkable production of electron–positron pairs. Indeed, the discrepancies found by Chao, regarding the predictions of the Klein-Nishina formula, were easily interpreted in terms of electron–positron pair creation by the γ rays and their ulterior annihilation.

We finish this account on the positron discovery with a phrase from Gordon Fraser in his book "Antimatter: The Ultimate Mirror", which we fully endorse: "*It is ironic that Dirac, the father of the positron, worked almost next door to the leading subnuclear laboratory in the world, and still the positron was discovered in California*".

Not only was this ironic but, the more one investigates about the positron discovery, the more one gets the impression that Carl Anderson was the "Chosen One" to accomplish this important mission that changed Physics forever. For if Skobeltzyn had had the initiative to upgrade his cloud chamber with much stronger magnetic fields he could have discovered the positron already in 1926; or if Auger had built his own cloud chamber correctly he could have made the discovery in early 1931; or if Blackett had discussed the existence of the anti-electron with Dirac and had reacted accordingly he could have made the discovery also in 1931 (they seemed to lack the ability to communicate with each other); or if Chao had had the initiative to try his experiments in a cloud chamber he would have seen positron tracks around 1929 (although having Millikan as supervisor, those observations would not had necessarily resulted in the positron discovery); not to mention all the other scientists, including Skobeltzyn, who missed the discovery in 1931–32 for not inspecting carefully the cloud chamber photographs which recorded their experiments with radioactive substances.

Chronology of the Positron Discovery

1911–23: Charles Wilson invents the cloud chamber and is the first person to photograph, at least until 1923, traces of cosmic rays, being unable to identify them as such, and unable to distinguish between electron and positron tracks because in the absence of a magnetic field they look identical.

1925: Dmitri Skobeltzyn, at Leningrad Polytechnical Institute, starts observing straight tracks of very energetic particles traversing his Wilson cloud chamber operated in a magnetic field. He realizes these particles must be part of the cosmic radiation. Since the tracks are not bent by his magnetic field it is impossible to distinguish the electric charge and thus between electrons and positrons. He publishes the first photograph of these straight tracks in 1927 and several more in 1929.

1927: Paul Dirac, at Cambridge University, writes the article *The Quantum Theory of the Electron,* in late 1927, being published February 1, 1928. There he presents the equation describing relativistic electrons, which has also solutions with negative energy.

1929: On a first attempt to make sense of the negative energy solutions, in 1929 Dirac writes the article *A Theory of electrons and protons,* which is published January 1, 1930. There he asserts that an electron with negative energy would behave as though it carries a positive charge, and he proposes that the proton could be the positively charged "electron". Robert Oppenheimer, at Caltech, reacts immediately against Dirac's proposal and shows that if the proton were the positively charged "electron", then the hydrogen atom as well as all ordinary matter would self-destruct in about 10^{-10} s.

1931: Dirac writes the article *Quantised Singularities in the Electromagnetic Field,* published September 1. There he proposes that the positively charged "electron" should be a new particle unknown to Physics with the same mass as an electron. He calls that particle "anti-electron" and argues that electron- anti-electron pairs could be created experimentally with γ rays. He also postulates the existence of the "anti-proton".

1931: From late summer, Carl Anderson at Caltech, investigating cosmic rays with a cloud chamber inside a very powerful magnetic field, makes hundreds of photos and realizes that many tracks which apparently are due to electrons, are bending the other way, indicating positive electric charge. He does not ascribe these tracks necessarily to protons, unlike Robert Millikan, who shows 11 of such photos at talks in Paris and Cambridge. Skobeltzyn reacts immediately against Millikan's interpretation, in a letter to Frederic Joliot-Curie, explaining that the positive charged particles in Anderson's photos, whose traces look very similar to electron traces, cannot possibly be protons.

1931: On December 19, a photograph entitled *Cosmic Rays Disrupt Atomic Hearts* is published in "The Science News-Letter", explaining some of the results of Anderson and showing the tracks of an ordinary electron and a positive charged particle, which is presented by the science journalist as "probably" (not surely) a proton. It was a positron.

1932: On August 2, Anderson takes the first confirmed photograph of a positron track, after having passed the "lead plate" test. Then he writes a brief report, *The Apparent Existence of Easily Deflectable Positives*, which is published September 9 in "Science".

1933: Patrick Blackett and Giuseppe Occhialini write the article *Some Photographs of the Tracks of Penetrating Radiation,* which is published March 3. They show for the first time the creation of electron–positron pairs from γ rays, which confirms that the positron of Anderson is the same particle as the anti-electron of Dirac.

1933: Anderson writes the article *The Positive Electron*, published March 15, with a complete account of his discovery. The editor indicates in the abstract: *"These particles will be called positrons"*, as suggested by Anderson as contraction of "positive electrons".

1933: As the results of Anderson, Blackett and Occhialini spread around, some other scientists, including Skobeltzyn, find positron traces in photographs from previous experiments recorded in cloud chambers in 1931–32; such traces had been either overlooked, or misinterpreted or mistaken for those of protons.

4.3   Antiprotons and Antineutrons

In the years following the discovery of the positron in 1932, several more particles together with their antiparticles were discovered in the cosmic rays. The first ones were the muons in 1936, followed more than 10 years later, in 1947, by the pions and kaons, which are composite hadrons called mesons, as was explained in Chap. 2 (Sect. 2.4.3, see also Fig. 2.4). However, it still took 22 years from the positron discovery for the existence of the most sought-after antiparticle—the *antiproton*—to be confirmed by experiment. Its discovery in 1955 did not take place in the traces of cosmic rays, though, but the antiproton was artificially produced in the most powerful particle accelerator at the time: the Bevatron, at the University of California in Berkeley. This was fortunate, not only for the sake of Physics itself but also because that machine had been constructed the year before with the specific aim of discovering the antiproton, predicted by Dirac in 1931, as this particle was very hard to find

Fig. 4.11 Emilio Segrè (1905—1989) (left) and Owen Chamberlain (1920—2006) (right) when they were awarded the 1959 Nobel Prize in Physics for the discovery of the antiproton. *Credit* The Nobel Foundation

in the cosmic radiation. So, failure to create the antiproton could have had unpredictable consequences for the Particle Physics community, at least in the USA.

To achieve this goal, the team led by Emilio Segrè and Owen Chamberlain (Fig. 4.11) collided proton beams against a copper target, reaching a total energy of 6.2 GeV, enough to create a proton-antiproton pair[6] in one out of five million collisions, approximately. Remarkably, in 1956, less than one year after the antiproton production had started, another team working at the Bevatron, led by Bruce Cork, was capable of using antiprotons to create *antineutrons* as well, but through $p\,\overline{p}$ collisions between proton and antiproton beams. Although the team used a very sophisticated experimental setting in order to detect the electrically uncharged antineutrons through their interactions with other particles, this discovery did not come as a big surprise as the stage for the antineutron appearance was perfectly set. Namely, due to electric charge conservation the by-products of the $p\,\overline{p}$ collisions had

---

[6]Since the internal structure of hadrons is very complex, the energy necessary to produce a proton-antiproton pair is much higher than the sum of their masses. In fact, an energy of about 6 GeV was necessary even though the proton mass is less than 1 GeV/c^2. See Chap. 6 for more details on particle production in accelerators.

to be neutral and, on the other hand, these collisions would involve plenty of antiquarks that would be available to form antineutrons.

Another important milestone in the production of antiprotons was carried out at CERN in 1978, when the Initial Cooling Experiment (ICE) was able to store antiprotons, keeping them circulating for 85 hours. This initiated the production of high energy antiprotons on a large scale in particle accelerators, in order to inject them into other larger and more powerful accelerators and make them collide with protons. For this purpose, a source of antiprotons was built in 1980 at CERN, the Antiproton Accumulator, which stored high energy antiprotons produced in the Proton Synchrotron (PS) for injection into the larger Super Proton Synchrotron (SPS). This was the first proton-antiproton collider ever built, although the first $p\,\overline{p}$ collisions which took place at CERN, in 1981, were produced in the Intersecting Storage Rings (ISR) shortly before starting on a large scale in the SPS accelerator.

Next, in 1982 the machine LEAR (Low Energy Antiproton Ring) was built at CERN and operated until 1996. It was a multipurpose machine (Fig. 4.12) that was supplied with a fraction of the antiprotons produced in the PS accelerator, and was mainly a decelerator, a cooler ring, and a storage ring. It opened many fields of research by providing, for the first time, a clean source of antiprotons with lower energies than those provided directly by the PS. Indeed, as many as 27 experiments were performed during the 14 years with LEAR's assistance, PS196 being the first experiment which succeeded in trapping antiprotons and keeping them for extended periods of up to 2 months. In the first years, the antiproton intensity delivered by LEAR was rather low so that it was mostly used for studying annihilations of antiprotons with nuclei (see Fig. 2.5). However, in the 1990s, LEAR delivered a record of one million antiprotons per second in one hour periods. In 1995 the experiment PS210, performed by a team led by Walter Oelert, achieved to create antihydrogen atoms by shooting a beam of antiprotons on xenon gas. The final account was the creation of nine antihydrogen atoms.... whose existence lasted only a few billionths of a second. But let us return to this experiment below, in Sect. 4.5, where we will address the creation of the very first antiatoms.

Meanwhile, in 1983, the production of high energy antiprotons also began on a large scale across the Atlantic, at the Fermi National Accelerator Laboratory, near Chicago and known as Fermilab. These antiprotons were produced to supply the $p\,\overline{p}$ collider named Tevatron, which became the most powerful particle accelerator in the world until 2009, and used of the order of 10^{13} antiprotons per year, at first, and around 10^{15} in its best years.

Fig. 4.12 The LEAR complex in 1983. It was a multipurpose machine supplied with a fraction of the antiprotons produced in the PS accelerator. It was mainly a decelerator, a cooler ring and a storage ring for antiprotons and it allowed to perform 27 different experiments. In 1995, the experiment PS210 carried out at the LEAR machine created nine atoms of antihydrogen for the first time in the world. *Credit* Courtesy of CERN

Furthermore, although it took a very long time, antiprotons were also detected in the cosmic rays, first by balloon-borne experiments and since 2006 also by satellite-based detectors. It was in 1979 when R. L. Golden, from the Physical Science Laboratory, Las Cruces, New Mexico (USA) and seven more collaborators published the article *Evidence for the Existence of Cosmic-Ray Antiprotons*, in "Physics Review Letters". They reported the observation of 46 antiprotons, from which 18 were expected to be atmospheric, that is, created by the action of cosmic rays colliding with the atoms in the atmosphere. We will come back to this in Sect. 4.6, where primordial versus secondary antimatter will be discussed.

4.4   Antinuclei

In 1965 two independent groups obtained the first atomic nucleus of antimatter beyond an antiproton (which is the nucleus of the antihydrogen atom $\overline{H}$). It was the *antideuteron*, consisting of one antiproton $\overline{p}$ and one antineutron $\overline{n}$, being the nucleus of the antideuterium atom, denoted as $\overline{D}$ or $^2\overline{H}$, which has never been observed. The production of the antideuteron was almost simultaneously achieved by a group led by Antonino Zichichi at CERN, and another group led by Leon Lederman at the Brookhaven National Laboratory in New York.

The next stable nucleus of antimatter, corresponding to *Antihelium-3*, $^3\overline{He}$, with two antiprotons and one antineutron, was first observed in 1971 in an experiment at the Protvino High Energy Institute, near Moscow, where aluminum nuclei were bombarded with protons. A total of five $^3\overline{He}$ nuclei were identified among 2.4×10^{11} particles that passed through the detector. Later on, the researchers improved the technique and bombarded nuclei with other nuclei.

The production of the following stable antinucleus, that of *Antihelium-4*, $^4\overline{He}$, also called $\overline{\alpha}$ (antialfa) particle, was only achieved in 2010 in the STAR experiment carried out at the Relativistic Heavy Ion Collider (RHIC) at Brookhaven. As suggested by its name, in this accelerator very heavy ions are collided with each other, and the objective is to reproduce the so-called quark-gluon soup, a plasma that only existed in the first instants of the primordial Universe composed of quarks, antiquarks and gluons. The STAR collaboration, with over 300 members representing 54 institutions in 12 countries, made use of gold–gold collisions in order to create the $^4\overline{He}$ nuclei and they had to sift through the debris of about one thousand million such collisions to finally identify 16 $^4\overline{He}$ nuclei. In the publication of these results in "Nature", in 2011, they reported 18 examples as they had two more from an earlier run.

After Antihelium-4, the next stable antimatter nucleus would be *Antilithium-6*, but its production rate in accelerators is expected to be over two million times less than for Antihelium-4. Therefore, the creation of Antilithium-6 nuclei is out of reach of our present accelerator technology and would probably require a major breakthrough. As for the detection of antinuclei in cosmic rays, only antiprotons have been detected, since 1979, but no other antinuclei, not even the antideuteron.

4.5   Antiatoms

It is unknown if antimatter atoms exist in nature somewhere in the Universe. Indeed, the only antiatoms we have known were created in laboratories, between 1995 and 2018, especially at the Antimatter Factory at CERN, where they will be produced again starting in August 2021. Let us have a closer look now at the first years of that undertaking.

As already explained, in 1995 the PS210 experiment was conducted at CERN in the LEAR ring, whose aim was to create antihydrogen atoms by shooting a beam of antiprotons on a fine jet of xenon gas. The leader of the research team, Walter Oelert (Fig. 4.13), hoped that the energy of the collisions would create enough electron–positron pairs so that a non-negligible fraction of positrons would be captured by antiprotons to form antihydrogen atoms $\overline{H}$. This was the third time that the team tried to create these antiatoms, having previously failed in 1993 and 1994.

Once the experiment was finished, from a thousand million initial antiprotons, about 23,000 events had a chance to correspond to antiatoms and had to be analyzed one by one. At first, a total of eleven atoms of $\overline{H}$ seemed to have been created in this experiment, but it turned out that two of the candidates were actually antineutrons, as the referee of "Physics Letters" suspected. So, the final account was the creation of nine antihydrogen atoms.... whose existence lasted only a few billionths of a second because they were immediately annihilated by the surrounding matter, generating a distinctive signal that could be measured and analyzed. A couple of years later, Fermilab announced the production of another hundred atoms of $\overline{H}$. However, like in the CERN experiment, these atoms were annihilated very quickly upon contact with the matter around, so that they could not be analyzed beyond identification.

Fig. 4.13 Walter Oelert in January 1996. He was the leader of the team which created the first atoms of antihydrogen in the LEAR ring at CERN, in 1995. *Credit* Courtesy of CERN

It should also be mentioned that much before the antihydrogen made its apparition, other kinds of exotic "atoms" involving antiprotons were created and analyzed in LEAR experiments. For example, the PS205 experiment *Laser Spectroscopy of Antiprotonic Helium Atoms* followed the discovery of antiprotonic helium atoms at KEK (Japan) in 1991. These exotic atoms are rather "atomcules" because they are hybrid between atoms and molecules, as they result from helium atoms with an electron replaced by an antiproton, and therefore they have two nuclei. Systematic studies of their properties were made at LEAR from 1991 to 1996.

The next task, to trap antihydrogen atoms $\overline{H}$ in a more stable manner, took only a few more years. Again, CERN took the lead by constructing the Antiproton Decelerator (AD), in the year 2000, in order to produce slow antiprotons with much lower energies than those provided by LEAR, resulting in speeds of one-tenth the maximum speed c. These antiprotons were then supplied to the various experiments that were emerging, installed in the enormous AD hall, which created antihydrogen atoms $\overline{H}$ in order to analyze whether they behave exactly like hydrogen atoms H. But we will continue this account in Chap. 8, where all those experiments will be briefly reviewed.

4.6   Primordial Versus Secondary Antimatter

It is crucial to differentiate between primordial and secondary antimatter. The former was created at the beginning of the Universe, in the first instants after the Big Bang, some 13,800 million years ago, and may well have disappeared in almost its entirety. Secondary antimatter, by contrast, is continually being created all over the Universe and in our environment; for instance, in collisions between particles. This is the case for cosmic rays impacting on atoms of the interstellar medium and atoms of our own atmosphere, giving rise to cascades of secondary radiation where matter and antimatter particles are found in almost equal numbers. In this respect, it should be noted that there are many astrophysical processes—especially neutron star pulsars and massive black holes—that release enormous amounts of antimatter into outer space, so they also contribute to the small fraction of antimatter found in the primary cosmic rays that reach the Earth—less than 1% of the total—consisting solely of antiprotons and positrons.

4.6.1 Primordial Antimatter

As argued in Chap. 7, it cannot be discarded that some primordial antimatter managed to survive the Great Annihilation against matter that took place just after they both were created. In that case, it could have given rise to large structures like antistars, antiplanets and even small antigalaxies in the actual Universe. Antistars would be indistinguishable from ordinary stars judging by their appearance, as they would emit essentially the same electromagnetic spectrum, but we still would be able to tell them apart because antistars would emit antineutrinos instead of neutrinos.

In any case, there is almost complete consensus in the scientific community that the dominance of matter over antimatter in our observable Universe is overwhelming. The relevant astronomical observations, which in principle could provide information about the existence of primordial antimatter in the Universe, belong to three areas: the *Cosmic Rays* , the *Diffuse Gamma Ray Background* (DGRB) and the *Cosmic Microwave Background* (CMB). We will have a close look at the Cosmic Rays in Chap. 5, and at the DGRB and CMB radiations in Chap. 7, Sect. 7.1.2. For this reason, in what follows we will only summarize the most relevant findings deduced from those astronomical observations.

First of all, it must be emphasized that only antimatter in the form of antiprotons and positrons has been found so far in the cosmic rays, not even antideuterons, even though the AMS-02 experiment, installed in the International Space Station, has collected more than 168 thousand million cosmic-ray events to date (November 2020).

Second, the observed positron flux in the cosmic rays has a remarkable excess for high energies with respect to the estimates of the secondary positron production in collisions of cosmic rays with atoms in the interstellar medium and other known astrophysical processes. But this excess does not suggest the existence of primordial positrons and can be explained in terms of very powerful positron sources originating in astronomical objects like neutron star pulsars and supermassive black holes. Alternatively, this excess could be a signal of dark matter annihilation.

Third, the observed antiproton flux has also an excess, but much less pronounced than for positrons. Again, this excess does not seem to require primordial antiprotons and could also be explained in terms of dark matter annihilation.

Fourth, the AMS-02 experiment has found about ten events which are candidates for nuclei of Antihelium-3 and Antihelium-4, although this possibility could only be confirmed around 2024, the earliest. If confirmed, according to some research groups, those antihelium nuclei very likely might be primordial and their very existence would provide evidence for antimatter dominated regions in the form of antigas clouds or antistars. The rate of anti-helium versus helium in our local Universe would be roughly one antihelium atom in a hundred million helium atoms, that is 10^{-8}.

Fifth, the *Diffuse Gamma Ray Background* does not show signals of gamma radiation from possible borders between domains of matter and domains of antimatter, where great amounts of γ rays could result from matter–anti-matter annihilation. Although the properties of those borders depend very much on the chosen cosmological models, the non-observation of gamma radiation boundaries between matter and antimatter domains leads to the conclusion that primordial antimatter, if it exists at all in the observable Universe, must be present in very small quantities compared to ordinary matter.

Sixth, the *Cosmic Microwave Background* shows no distortions of non-thermal origin superimposed to the thermal spectrum of 2.7 K that could be interpreted as signals of annihilation between matter and antimatter domains, leading to the same conclusion as in the previous paragraph.

4.6.2 Secondary Antimatter

Now we will consider the production of secondary antimatter in nature, that is, of positrons and antiprotons. No other kind of antimatter has ever been observed outside laboratories, although this might change in the next few years due to the observations of the cosmic rays in space, in particular by the AMS-02 experiment, as explained above.

Positrons e^+.

Besides the collisions of cosmic rays with the atoms of gas and dust in the interstellar medium, the nuclear reactions within stars and many other astro-physical processes also produce enormous quantities of antimatter, mainly in the form of positrons. However, the positrons produced in the nuclear reactions inside stars annihilate very quickly with the electrons in the proton-electron plasma where they are born. As a result, stars do not contribute substantially to the positron abundance in interstellar space since no positrons manage to escape from the stellar cores into outer space (see Fig. 7.1 in Chap. 7). However, collisions of protons with energies above 300 MeV in dense layers of the stellar atmospheres do create positrons via pions, through the decays: $\pi^+ \rightarrow \mu^+ \rightarrow e^+$, (Eq. 2.3 in Chap. 2). These positrons have been observed in solar flares since the 1970s through their annihilation radiation—511 keV γ rays—and also directly since 2013.

On the other side of the scale, positrons are also created in the atomic nuclei of certain radioactive substances, such as the potassium isotope ^{40}K, which is abundant in walls, bananas and in our own bones. To be precise, the ^{40}K isotope emits 89.28% of the time one electron plus one antineutrino, that is, β radiation; 10.72% of the time it emits gamma radiation plus a neutrino; and 0.001% of the time it emits a positron plus a neutrino, that is, β^+ radiation. The positrons emitted by our bones, about 3800 per day for a person weighing 70 kg, do not leave the body, as they are annihilated by the electrons they encounter on their way, their free path being barely one millimeter. In the case of bananas, however, since they also have ^{40}K in the skin, about 15 positrons escape into the environment every 24 h. Hence, bananas do emit antimatter, and the same can be said of ceramics and walls, since lime and cement also contain ^{40}K.

The major sources of positrons seem to be neutron star pulsars and the regions near massive black holes. This suspicion dates back to 1998 when huge jets of electron–positron plasma were identified for the first time close to enormous black holes which created them in four active galactic nuclei (AGN)[7] (see Fig. 3.22 in Chap. 3). Then, in 2008, a research group discovered that a large population of binary systems containing black holes and neutron stars was at the origin of a giant cloud of electron–positron plasma surrounding the galactic center of our Milky Way. This cloud, which had been identified in the 1970s through the γ ray emission from the $e^+ e^-$ annihilation, extends for 10,000 light years, approximately.

At present, it is believed that the electron–positron plasma is actually the main component of most relativistic jets, emitted by neutron star pulsars (Fig. 4.14) and massive black holes, although by means of not well understood mechanisms. Their lengths can reach thousands or even hundreds of thousands of light years. This also implies that neutron star pulsars and massive black holes are the prime suspects behind the generation of Ultra-High Energy Cosmic Rays, as will be discussed in the next chapter.

---

[7]Active galactic nuclei (AGN) are at the center of galaxies and quasars and have a much higher luminosity than average. In addition, the analysis of that excessive luminosity, which has been observed at all frequencies of the electromagnetic spectrum, shows that it is not produced by stars and is thought to result from the accretion of matter by a supermassive black hole.

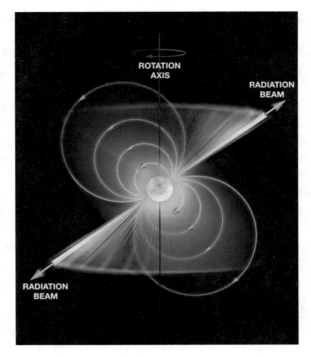

Fig. 4.14 Artist's illustration of a neutron star pulsar. In the center is the neutron star showing its magnetic poles from which the magnetic field lines and the radiation jets emerge. In vertical is the axis of rotation of the star. Since the magnetic poles and the rotation poles do not coincide (unlike here on Earth), the jets sweep through space and can only be detected if they point directly towards the observer. *Credit* B. Saxton, NRAO/NSF (National Radio Astronomy Observatory and National Science Foundation)

Neutron Stars and Black Holes in a Nutshell

Neutron stars and black holes represent the final stages of the evolution of stars, together with white dwarf stars, once they have exhausted their nuclear fuel. The less massive stars, until around 10 solar masses, become white dwarfs (see more details in Chap. 3, Sect. 3.5.1). If the star has a mass between 10 and 25 solar masses, then it eventually explodes as a supernova and its core ends up as a neutron star, with a radius of the order of 10 km and a mass between 1.44 and 2.16 solar masses. It is made of neutrons, essentially, which are created via electron capture by the protons, the latter "becoming" neutrons with the aid of the W bosons of the weak interactions (process known as *inverse beta decay*). Neutron stars are supported against further gravitational collapse by the *neutron degeneracy pressure* together with strong repulsive nuclear forces. There are some fascinating facts about neutron stars. One is their shockingly strong magnetic fields, of unclear origin, which have values between 10^8 and 10^{15} times stronger than Earth's magnetic field, and several orders of

Fig. 4.15 Jocelyn Bell in 1967, the same year she discovered neutron star pulsars while she was a research assistant searching for quasars under the supervision of Antony Hewish, her thesis advisor at Cambridge University. She helped build a large radio telescope with which she made her discovery. Although she had a hard time convincing Hewish that the pulsating radio sources she had found were from outer space, and not from terrestrial devices—as he believed—Hewish received the Nobel Prize in 1974 for this discovery, but not Jocelyn. Since then she has received many awards and honors, like the *Special Breakthrough Prize* In *Fundamental Physics, in 2018. Credit* Author: Roger Haworth. Courtesy of Wikimedia Commons (Creat. Comm. 2.0 license)

magnitude stronger than any magnetic field observed so far in other known astrophysical objects. Another fascinating fact is their extremely rapid spinning velocity around their axis of rotation, to the extent that the fastest spinning neutron star detected to date, known as PSR J1748-2446ad, rotates at a rate of 716 times a second.

Most neutron stars emit powerful jets of electromagnetic radiation from the regions near their magnetic poles, especially in the range of the radio frequencies, but they also emit in X-rays and γ rays, and not necessarily through the magnetic poles. If these poles are not aligned with the rotation poles, which is the general case, then the jets sweep through space, like a lighthouse. From a very far distance, the jets can only be detected if they point directly at us, and they are perceived as pulses, like blinking dots if they were emitting in the optical visible light. Such neutron stars were called pulsars by a science reporter of The Daily Telegraph, who shortened 'pulsating radio source' to pulsar. They were discovered by Jocelyn Bell (Fig. 4.15) in 1967, while she was a graduate student in Cambridge University searching for quasars. Pulsars provided the first evidence for the existence of neutron stars, even though their existence had been conjectured already shortly after the discovery of the neutron in 1932.

If the star has a mass beyond 25 solar masses, then it also eventually explodes as a supernova, but its remnant becomes a black hole as no physical processes can stop the gravitational collapse crushing it. They are called black holes because, according to General Relativity, they are surrounded by a boundary—the *event horizon*—from which nothing can escape, not even light. Hence, black holes are really black. Their presence can be inferred, however, through their interaction with other matter and radiation and through their gravitational lensing deforming the images of galaxies and other astrophysical objects. Black holes grow by swallowing everything within their reach, including whole stars, and by merging with other black holes. In this way, supermassive black holes can form, with millions and even thousands of millions of solar masses. Indeed, it is believed that supermassive black holes exist in the centers of most galaxies, and surely in the centers of all active galactic nuclei (AGN), such as quasars. In the center of our Milky Way there is a supermassive black hole of "only" 4.3 million solar masses.

It should also be noted that primordial black holes are conjectured to exist. They would have been created in the first moments of existence of the Universe, much earlier than atoms. If so, they might contribute substantially to the dark matter of the Universe, and they could be at the origin of many (or all) supermassive black holes.

Furthermore, a recent study appeared in December 2019, using 10 years of γ-ray data from the Fermi Large Area Gamma-ray Space Telescope (Fermi LAT), presents sound evidence that the excess of positrons observed near Earth is coming from pulsars in the Milky Way. The study was done by Mattia Di Mauro (NASA), Silvia Manconi and Fiorenza Donato, from the University of Torino (Italy), and was published in the article *Detection of a γ-ray halo around Geminga with the Fermi-LAT data and implications for the positron flux,* in "Physics Review D".

The data analysis was centered on the Geminga pulsar, 815 light years from us in the Gemini constellation, around which an enormous γ-ray halo was discovered, as seen in Fig. 4.16. This halo is believed to form due to the action of the energetic electrons and positrons released by the pulsar from its magnetic equator, rather than from the magnetic poles. The reason is that Geminga's magnetic field is so powerful that a great amount of its

Fig. 4.16 These images, artificially colored, show a region of the sky centered on the pulsar Geminga as seen in γ rays by the Fermi Large Area Gamma Ray Telescope (Fermi LAT). On the left, the total number of γ rays detected by Fermi LAT at energies from 8 to 1000 GeV over the past decade. Brighter colors correspond to greater numbers of γ rays. On the right, by removing all bright sources, astronomers discovered the Geminga's extended γ-ray halo. *Credit* Courtesy of NASA and Fermi LAT Collaboration

high energy photons creates electron–positron pairs. Some of those electrons and positrons annihilate with one another, but many others collide with the surrounding starlight, transferring some of their energy to these low energy photons,[8] which eventually reach much higher energies in the γ-ray spectrum. The team concluded that the pulsar Geminga alone could be responsible for as much as 20% of the high energy positrons detected by the AMS-02 experiment.

Curiously, Geminga was the first example of an unidentified mysterious γ-ray source which could not be associated with any objects. Its nature was unknown for about 20 years until, in 1992, it was identified to be the first example of a pulsar whose radio emissions we cannot observe from Earth because they are not pointing in our direction, as explained in Fig. 4.17, where one can see Geminga as seen in X-rays. Besides γ rays, for which it is a very bright (non-pulsating) source, Geminga is also visible, although with some difficulty, in X-rays and in the optical spectrum with an apparent magnitude of 25.5.

---

[8]This effect is called Inverse Compton Scattering.

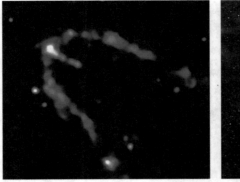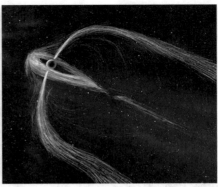

Fig. 4.17 On the left, X-ray image (colored in blue and purple) of the Geminga pulsar and its wind nebula as observed by the Chandra X-ray Observatory. Geminga was first detected as a bright γ-ray source in 1972, and identified as a neutron star pulsar in 1992. It has a period of 0.237 s, which means that it turns around its axis about four times per second. Its radius is around 10 km, and it could have originated in a supernova explosion some 342,000 years ago. The illustration on the right depicts more details of the astronomers interpretation of the X-ray image. The long narrow trails on both sides of the little Geminga represent radio jets emanating from its poles (its magnetic and rotation poles are very close to each other). These jets do not point towards the Earth, so we cannot detect Geminga's radio emissions. There is also a toroidal disk-shaped region of γ ray emission spreading from the pulsar's magnetic equator. This region and the jets are pushed back as the pulsar moves through the Galaxy. *Credit* NASA/Penn State University (PSU); Illustration: Nahks Tr'Ehnl (PSU)

Back to Earth, in 2010 the Fermi Gamma-ray Space Telescope accidentally spotted thunderstorms producing beams of electron–positron pairs, which were discovered above the thunderstorm clouds. They are produced in flashes of γ rays created by electrons, accelerated by the strong electric fields, as they collide with the atoms and molecules they encounter. This finding implies that very probably many planets have thunderstorms that can create positrons and then launch them into space.

Another source of positrons could be the decay of dark matter particles or their annihilation between themselves creating positrons as by-products. This hypothesis has been proposed by several research groups, especially after the space-based PAMELA experiment reported in 2008 the observation of a positron excess for energies above 10 GeV. This excess is with respect to the expected abundance of positrons produced by the collisions of cosmic rays with the atoms of gas and dust in the interstellar medium. More recently, the AMS experiment has improved substantially the results of PAMELA regarding the positron excess, as will be explained in detail in Chap. 5.

Antiprotons $\overline{p}$

Antiprotons are also detected in the cosmic rays, as explained in Sect. 4.3. They are mainly produced in the collisions of cosmic-ray protons with atoms in the interstellar clouds, as well as atoms in our atmosphere, according to the process:

$$p + A \rightarrow p + \overline{p} + p + A, \tag{4.3}$$

where A is the atomic nucleus. Therefore, antiprotons can be found as part of the primary cosmic rays that penetrate our atmosphere, but also as part of the secondary radiation produced by the collisions of the primary cosmic rays with the atmospheric atoms. In this case they are called atmospheric antiprotons.

Indeed, in the 1979 article *Evidence for the Existence of Cosmic-Ray Antiprotons*, by R.L. Golden and seven collaborators, where the discovery of antiprotons in cosmic rays was announced, the researchers reported the observation of 46 antiprotons, from which 18 were expected to be atmospheric. Although they found so few antiprotons, they could estimate the ratio 5×10^{-4} of antiprotons over protons, not very far from the most updated mean value of 1.81×10^{-4} obtained by the AMS collaboration in 2016. However, the authors concluded that that ratio was consistent with the secondary production of antiprotons in the stellar medium, which is not the case since even the lower value found by the AMS experiment seems to be above the expected secondary production of antiprotons in the stellar medium. In fact, since 2015 various groups have identified an excess of antiprotons in the fluxes of cosmic rays observed by the PAMELA and AMS experiments. We will come back to this issue in Chapter 5, Sect. 5.6.3.

Another source of antiprotons, like for positrons, could be the decay of dark matter particles or their annihilation between themselves. This is precisely the proposal of the groups just mentioned that identify an excess of antiprotons in the fluxes of cosmic rays.

Interestingly, it turns out that antiprotons in outer space can propagate through very long distances, their survival time in our Galaxy being between one and 10 million years until they annihilate themselves with matter. But antiprotons can also be captured and confined by magnetic fields (and black holes!) very soon after their creation. This the case of a large number of antiprotons trapped in the inner Van Allen radiation belt surrounding Earth, which was discovered by the PAMELA experiment in 2011. More details about this in Chap. 5.

5

Cosmic Rays

5.1   The Cosmic Ray Pioneers

At the beginning of the twentieth century, together with the recently discovered radioactivity, some curious electrostatic phenomena in the atmosphere were also investigated. Something unknown was pulling, or adding, electrons to the air molecules, so that they became electrically charged producing the ionization of the air. Not surprisingly, the idea that the air ionization was due to the presence of radioactive materials on the Earth's surface soon spread. In 1910, a Jesuit priest named Theodor Wulf was probably the first physicist who performed an experiment in order to observe the relation between the air ionization and height, to test such hypothesis. Armed with an electrometer he had built himself - an instrument for measuring electric charge - he climbed to the top of the Eiffel Tower in Paris, which is 330 m high. There he found that the air presented less ionization than the air at the bottom, but still a much higher ionization than expected under the assumption of a purely terrestrial origin. After four days of experiments, he published an article in the journal "Physikalische Zeitschrift" with an account of his observations.

Shortly after, from 1911 to 1913, Victor Hess (Fig. 5.1) made ten balloon experiments over Austria and Germany to investigate the number of atmospheric ions present at different altitudes. The most famous of these experiments took place on 7 August 1912. The balloon Bohemia (Böhmen) was launched at 6:12 from Aussig an der Elbe (now in the Czech Republic), rising to 5,350 m and landing six hours later close to the village Pieskow, 50 km east of Berlin. The measurements of Hess showed that the ionization of the air first presents a decrease during the ascend, in accord with Wulf

© Springer Nature Switzerland AG 2021
B. Gato-Rivera, *Antimatter*,
https://doi.org/10.1007/978-3-030-67791-6_5

Fig. 5.1 Victor Hess (1883–1964) landing from one of his balloon flights in Austria, in 1912. *Credit* Instituto Argentino de Radioastronomía and CERN

results, but then a strong increase above 1,400 m. Therefore, Hess proved that the air ionization increases with altitude and, consequently, it could not be due to radiation from the surface but to radiation coming either from outer space or from the upper layers of the atmosphere. In fact, he did not exclude that radiation could be created somehow by the electricity or anything else present there. Either way, this radiation, which Hess named with the German term *Höhenstrahlung* (altitude radiation) was bombarding the Earth in all directions. In addition, five of the flights were made at night and one during a solar eclipse, finding that there were no appreciable differences between the quantities of diurnal and nocturnal ions, and therefore these radiations did not come from the Sun. He published an article in "Physikalische Zeitschrift" with the results obtained during his balloon campaigns in 1912. Three decades later, he summarized the highlights of that campaign in the article *The discovery of cosmic radiation*, published in 1940 in the journal "Thought: Fordham University Quarterly". He said:

> The only possible way to interpret my experimental findings was to conclude to the existence of a hitherto unknown and very penetrating radiation, coming mainly from above and being most probably of extra-terrestrial (cosmic)

origin.... During a solar eclipse in 1912, I found no change of ionization; from this I concluded that this cosmic radiation could not originate in the Sun itself, at least not as an electromagnetic radiation....

The air ionization measurements of Hess were extended to higher altitudes up to about 9,000 m by Werner Kolhörster in 1913, by which he confirmed the increase of the air ionization with altitude. Around 20 years later, in 1934, a balloon was launched with two collaborators of Kolhörster in order to perform measurements up to altitudes of 12,000 m. They brought a breathing apparatus and oxygen supply for four hours. During the flight, unfortunately, an accident happened, and the two physicists died. In the meantime, however, Robert Millikan led a number of experiments in USA in the summers of 1922, 1923 and 1925 in balloons, airplanes and mountain peaks, which confirmed, in his own words, *"a definite variation of the penetrating radiation with altitude alone"* and he coined the name *cosmic rays*.

Meanwhile, in 1923, the Wilson cloud chambers reached their final stage for being used in research tasks. In the same year, Dmitri Skobeltzyn (Fig. 5.2), working at the Leningrad Polytechnic Institute (USSR), was one of the first scientists to construct and use such a device. He was investigating the scattering of γ rays[1] obtained from a radioactive source attached at the internal wall of the chamber and he wanted to see the tracks of the recoil electrons from the gas molecules hit by the γ rays. But there was a problem that was messing up the images; it turned out that the γ rays were also ejecting electrons from the walls of the chamber. To get rid of the unwanted electrons hanging around, in 1924 he decided to place the chamber in a magnetic field (see Fig. B1 in App. B). As a result, Skobeltzyn became the first scientist ever to observe cloud chamber traces of charged particles curving under the influence of a magnetic field.

Skobeltzyn soon realized that the analysis of the curvature of these tracks could give important information about the particles, such as their energies and electric charges. Then, in 1925 he discovered tracks left by particles not related to the γ rays from the radioactive source and with very high energy, as deduced from the fact that they did not deflect appreciably under the magnetic field. Indeed, the tracks were so straight that it was not possible to tell the electric charge of the corresponding particles, although all the experts, including Skobeltzyn, believed they were just electrons (often referred to as

---

[1]Skobeltzyn (also spelled Skobeltsyn and Skobelzyn) was investigating the so-called Compton effect, which had just been discovered by Arthur Compton in 1923. It consists of the scattering of an energetic photon (X-ray or γ-ray) by an electron of an atom, resulting in a decrease of the photon energy and momentum. This was the definitive proof that electromagnetic radiation consists of a stream of particles (photons). Compton received the 1927 Nobel Prize in Physics for this discovery.

Fig. 5.2 Dmitri Skobeltzyn (1892–1990) was a Soviet physicist, both theorist and experimentalist. He also worked, in 1929–1931, at the Curie Laboratory in Paris. He was a pioneer in Nuclear Physics, High Energy Physics as well as in Cosmic Rays, being the first physicist to put a Wilson chamber in a magnetic field (in 1924) and the first to see tracks left by high energy particles from the cosmic radiation. These were straight tracks in which it was not possible to distinguish the electric charge of the particles. He was also the founder and first Director of the Institute of Nuclear Physics in Moscow State University (USSR), founded in 1946, which in 1993 was named after him. *Credit* Soviet Academy of Sciences

β particles in those days). During 1925 and 1926 he took more than 600 photographs in a stronger magnetic field from 1500 to 2000 gauss (Fig. B2 in App. B), from which the straight tracks of the high energy particles showed up in 27 images.

His first photograph of a cosmic-ray track was published in 1927 in an article in the journal "Zeitschrift für Physik" (Fig. B3 in App. B), where he described such traces as tracks of *"particles of unknown origin with extremely high energy, certainly more than 20 MeV"*. In further work Skobeltzyn showed that the ionization rate of these particles was consistent with that of the cosmic rays, whose energies were unknown. Therefore, Skobeltzyn's experiments provided sound evidence that the high energy particles he had detected were part of the cosmic radiation discovered by Victor Hess. Moreover, careful analysis of some tracks showed that some particles shared a common origin, as if they had emerged from the interaction of a cosmic ray with an atom. In other words, Skobeltzyn had observed a cosmic ray shower for the first time. A full report of all those findings followed in 1929, also in "Zeitschrift für Physik". However, he had already disclosed the most relevant of these results in 1928 in a conference at the Cavendish Laboratory in Cambridge, as he explained in his 1980 contribution for the book "The Birth of Particle Physics":

> But most of the pertinent facts and photographs were presented by me earlier in the course of a discussion at an informal conference "On γ and β Ray Problems", held at Cambridge, July 23 to 27, 1928, under the patronage of Ernest Rutherford. One session of the conference was scheduled for a discussion on β-ray problems (it seems that no paper was read at that session). During that discussion, I demonstrated a collection of photographs of cosmic ray tracks, and, I dare say, it produced some impression on the audience.

Indeed, the images of straight tracks from cosmic-ray particles, obtained by Skobeltzyn in his cloud chamber operated in a magnetic field, were the spark that ignited the cosmic ray research using cloud chambers. The race started only two years later, with Patrick Blackett and Giuseppe Occhialini at the Cavendish Laboratory in Cambridge (UK), on one side, and Robert Millikan and his student Carl Anderson at Caltech, in California (USA), on the other side. But this was the subject of the previous chapter.

5.2   Cascades of Cosmic Rays

Cosmic rays can reach very high energies, although they arrive with the full spectrum of possible energies which cover more than 20 orders of magnitude. This provides them with a high penetration power in the atmosphere and, in addition, being equipped with electric charge, they also have a high ionization power when they collide with the air molecules. Remarkably, it took 35 years for the observations of Hess to be properly interpreted. It happened that he and several other physicists, including Millikan, believed that cosmic rays consisted probably of highly energetic γ rays coming from outer space and giving rise to electromagnetic cascades in the atmosphere.

Fig. 5.3 Pierre Auger (1899 - 1993) in the 1930s. He was one of the discoverers of the extensive cascades of cosmic rays, and he carried out the first systematic study of their structure and properties in 1939. The Pierre Auger Observatory, in Argentina, and since 2003 the largest observatory in the world for the study of cosmic rays, was named in his honor. *Credit* Courtesy of the Pierre Auger Observatory

In fact, in 1938 Pierre Auger (Fig. 5.3) and his collaborators at the Ecole Normale Supérieure, in Paris, discovered that cosmic rays came precisely in the form of cascades of particles that spread widely. This was the simplest explanation for the coincidences they had observed between the signals detected in Geiger counters[2] separated by distances of up to 300 m (in a horizontal plane). Curiously, only four years earlier, Bruno Rossi had also discovered this phenomenon while he was investigating east–west asymmetries of cosmic rays in Eritrea, but he did not provide any information about the distances between his Geiger counters nor did he pursue this research any further. In his own words (in an article written in an Italian journal) *"Unfortunately, I did not have time to study this phenomenon more closely"*.

Even so, the explanation that cosmic radiation was due to γ rays from outer space was not correct. Cosmic ray cascades do occur when particles from

---

[2]Geiger counters are instruments used to detect and measure radiation from both electrically charged particles and γ rays.

Fig. 5.4 Artist's illustration of primary cosmic rays impacting on Earth's atmosphere and giving rise to cascades of secondary radiation. *Credit* Asimmetrie/INFN

outer space collide with atoms in the atmosphere (Fig. 5.4), but the cosmic-ray primaries that initiate the cascades are mainly charged particles, not γ-ray photons, even though electromagnetic cascades also exist. The cascades of secondary radiation are formed when the cosmic-ray primaries penetrate into the atmosphere at high speeds. Then the collisions with the atoms can reach very high energies which create new particles (matter and antimatter particles half and half, roughly). Among these there are often heavy particles which decay producing lighter particles that can also decay or collide with other atoms, and so on. Now, the nature of the particles that compose the cascades depend on the nature of the cosmic-ray primaries initiating them. Therefore, these can be identified by analyzing the secondary radiation. Let us see more about this.

In 1947 the so-called Bristol group discovered the pions π^- and π^+ in the cosmic radiation,[3] together with their rapid decays into muons μ^- and antimuons μ^+, simply called muons (negative and positive). Those decays can be seen in Chap. 2, Eq. (2.3) and Fig. 2.5. The muons had been discovered in 1936 by Carl Anderson and Seth Neddermeyer with their magnet cloud chamber (Fig. 4.6), as explained in Chap. 4, but they were not fully

---

[3]They did this using photographic emulsion plates in high altitude places such as mountains and planes.

understood until the pions were discovered and analyzed.[4] Since the muons have a mass 207 times larger than the electron, the braking capacity of the atmosphere on these particles is much lower than on the electrons. Furthermore, since those muons move at velocities close to the speed limit c, the relativistic dilatation of their average lifetime is significant. Consequently, the muons produced in the cascades of cosmic radiation are much more penetrating into the atmosphere than the electrons and, in fact, they turned out to be the main component of the radiation observed by Victor Hess and the other cosmic ray pioneers.

Moreover, the proposal that the primary cosmic rays that initiate the cascades are γ rays had to be discarded when the pions were discovered. The reason is that pions cannot be produced in electromagnetic cascades, consisting essentially of electrons, positrons, and more photons. It was finally concluded that, although such γ-ray cascades also exist, the vast majority of them are of hadronic type, produced by the constant bombardment on Earth of atomic nuclei, especially protons and α particles. These nuclei constitute 98% of the primary cosmic rays endowed with electric charge (of which 90% are protons and 8% are α particles), the remaining 2% being electrons and more massive nuclei, as well as a small amount of antimatter in the form of positrons and antiprotons. Now, it should be noted that, nowadays, the name "cosmic rays" refers exclusively to the primaries with electric charge; the other high energy visitors from outer space are γ rays and neutrinos, and they are named accordingly. But, while γ rays give rise to electromagnetic cascades of secondary radiation, neutrinos do not generate any cascades because they interact very rarely with the atoms in the atmosphere (not even with the Earth itself).

The primary cosmic rays penetrate into the atmosphere and collide with a proton or a neutron of an atomic nucleus (mainly of nitrogen or oxygen). Then, if their energy is high enough the production of secondary radiation begins, as we have discussed. The number of secondary particles created in a cascade depends strongly on the energy of the primary cosmic-ray particle and can range from a few thousands to millions or thousands of millions of particles, and even more.

As shown in Fig. 5.5, there is a hadronic component in the center of the cascades, which corresponds to the remnants of the atmospheric nucleus

---

[4]This seems kind of paradoxical as muons are elementary particles while pions are hadrons composed of a quark and an antiquark. However, the muon was conjectured to be the mediator of the strong forces between the nucleons (protons and neutrons), postulated by Hideki Yukawa in 1935, and shortly after the discovery of the pions it became clear that the latter were precisely those mediators.

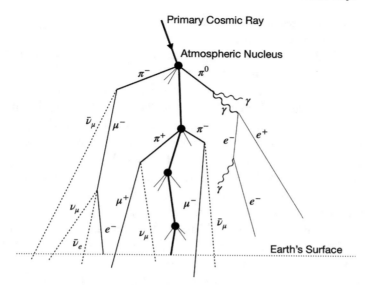

Fig. 5.5 Cascade of Cosmic Rays, where matter and antimatter particles are found in equal numbers, approximately. The majority of photons, electrons and positrons are absorbed in the atmosphere and only the most energetic ones reach the Earth's surface. The most energetic muons μ^- and antimuons μ^+ can penetrate hundreds of meters of rock before decaying, while neutrinos ν and antineutrinos $\bar{\nu}$ traverse the Earth unperturbed

with which the primary cosmic ray collided. There is also an electromagnetic component of electrons, positrons, and photons (e^-, e^+ and γ) that reaches the ground only partially since it is mainly absorbed by the atmosphere. Moreover, there are large numbers of muons (μ^-, μ^+), together with even larger numbers of electron neutrinos and antineutrinos (ν_e, $\bar{\nu}_e$) as well as muon neutrinos and antineutrinos (ν_μ, $\bar{\nu}_\mu$). As explained in Chap. 2, neutrinos and antineutrinos traverse the Earth without almost any interactions, continuing their journey undisturbed.

The muons μ^- and μ^+, despite being unstable and decaying spontaneously in just microseconds, often reach the Earth's surface due to their high penetrating power. In fact, it is estimated that the flow of muons that reach sea level is around one per cm^2 every minute. Moreover, the most energetic muons can pass through hundreds of meters of rock and kilometers of water before finally decaying, producing an electron e^- (from μ^-) or a positron e^+ (from μ^+). This means that muons can enter in our houses and buildings easily, which is why Anderson and Neddermeyer could discover them in the cloud chamber at the laboratory. This fact also suggests that many electrons and positrons recorded in cloud chambers were probably just created by the decay of very energetic muons.

The penetrating capacity of the most energetic muons in the cosmic ray cascades has recently been exploited to study the interior of the Great Pyramid of Cheops, in Egypt, where the international collaboration *Scan Pyramids* has detected a hidden chamber 30 m long. The detection method is very simple since it is based on measuring the flow of muons through the area of the pyramid under study and compare it with the flow that would reach the detectors if that area were made of solid rock. If more muons arrive than when they pass through the rock, this means that there is a cavity, because the muons are absorbed less through the air than through the rock, obviously. This technique was used for the first time by Luis Walter Álvarez and his team, in 1968, to search for hidden chambers in the Kefren pyramid, the second largest after the Great Pyramid. They explored 19% of the total volume of the pyramid and no hidden chambers were found.

We see then that atmospheric muons and antimuons are part of the natural radiation to which we are exposed, and not only in the open air but also inside our buildings. This has an impact on living beings and their evolution (although most experts think that it is unimportant), since the bombardment of biological tissue with highly energetic particles can result in DNA mutations as well as cell damage. As for the electrons and positrons in the cascades, although most of them are absorbed by the atmosphere, they reach the Earth's surface in quantities between 10 and 100 times larger than the muons, depending on the energy of the primary cosmic ray.

The radiation of very energetic charged particles do have consequences for the performance of electric instruments and electronic devices, as is well known for more than one and a half century; in particular in relation with the solar storms in which gigantic amounts of electromagnetic energy and charged particles (solar wind) are released by the Sun. The strongest solar storm ever recorded, known as the *Carrington Event*, took place on September 1–2, 1859. Fortunately, at that time electricity-based technology was very minor, so that not much damage could be produced. Now, in July 2012, some satellites discovered that another very powerful solar storm had just missed Earth; otherwise it would have caused lasting catastrophic effects due to our heavily electricity-dependent technology (and lifestyle).

Before we close this section, let us say a few words about the experiment CLOUD (*Cosmic Leaving Outdoor Droplets*), being conducted at CERN since 2009. The question to be answered is whether there is a relation between cosmic rays and atmospheric aerosol and cloud formation, as well as their implications for air pollution and climate. The results of these investigations have recently revealed a new mechanism that drives winter smog episodes in cities, published in "Nature" in May 2020. As for climate implications, their results seem to indicate that the effect of cosmic rays on aerosol and cloud formation is insufficient to attribute the present climate change to the fluctuations of the cosmic ray intensity modulated by changes in the solar activity and changes in the Earth's magnetic field.

5.3 Ultra-High Energy Cosmic Rays

One of the most amazing aspects of cosmic rays is the exceedingly high energies they can reach. The research carried out by Pierre Auger already pointed out that most of the large cascades discovered by his research team, also known as *air showers*, had been generated by primary rays with energies slightly above 10^{15} eV (but of this order of magnitude). He estimated these energies by assuming that the corresponding air showers consisted of a million particles, approximately, each carrying 10^8 eV of energy; and a factor 10 would take into account the energy loss through the atmosphere.

These results were totally spectacular, since most energetic particles that had been observed so far had energies several orders of magnitude lower, up to 10^8 eV for some cosmic rays registered in the cloud chambers. Nevertheless, the cascades discovered by Auger were not the most typical ones, as only a small fraction of the primary cosmic rays reach such high energies. In fact, they arrive with the full range of possible energies, as we already pointed out. But later even more energetic cosmic rays were detected, and the cosmic rays with energies of the order 10^{18} eV and beyond were dubbed *Ultra-High Energy Cosmic Rays* (UHECRs).

Furthermore, in 1963 John Linsley (Fig. 5.6) observed a primary cosmic ray with an energy around 10^{20} eV when analyzing a cascade in a facility in New Mexico (USA) called Volcano Ranch, which counted with an array of particle detectors. This produced quite a commotion, both in the community of Cosmic Ray Physics and in that of Particle Physics, because even today the physical mechanism capable of accelerating particles to such energies is totally unknown despite the abundance of very energetic astrophysical processes in the Universe. The possible sources of UHECRs which are considered include γ ray bursts as well as neutron star pulsars and mergers, but especially processes in active galactic nuclei (AGN) and in starburst galaxies. The AGN most probably contain a supermassive black hole, whose mass reaches between a million and a thousand million times the mass of the Sun, while the starburst galaxies have an exceptionally high rate of star formation (see Fig. 5.7).

Fig. 5.6 John Linsley (1925–2002) performed pioneering research on ultra-high energy cosmic rays. Passport photograph in 1963. *Credit* USA Government, Courtesy of Wikimedia Commons

Fig. 5.7 Possible sources of ultra-high energy cosmic rays (UHECR). On the left the Antennae Galaxies, a pair of interacting galaxies (NGC 4038 and NGC 4039) in collision, in the constellation Corvus, which provide an example of starburst galaxies. *Credit* Courtesy of ESA/Hubble and NASA. On the right, Centaurus A (NGC 5128), a galaxy in the Centaurus constellation with an active nucleus (AGN). *Credit* Courtesy of ESO/WFI and NASA

For comparison, the most powerful particle accelerator that has existed, the LHC at CERN, reached a maximum energy of $13 \, \text{TeV} = 1.3 \times 10^{13}$ eV; that is, seven orders of magnitude lower than the energies of the more energetic cosmic rays. However, it is not correct to equate the energy of a cosmic-ray proton, when colliding against an atmospheric proton, with the total energy in a head-on collision between two protons, as in the $p \, p$ collisions in the LHC. The reason is that one has to take into account relativistic effects[5] and, in fact, one of those collisions at 13 TeV is equivalent to an impact of a 10^{17} eV cosmic-ray proton against a stationary atmospheric proton, which is still 1000 times less energetic than 10^{20} eV.

The observation of Linsley prompted the construction of other facilities for detecting high energy cosmic rays, so that they could be studied and the frequency of their arrival on Earth could be analyzed as a function of their energy. It happens that the flow of high energy cosmic rays that enter the atmosphere is extremely low. For energies slightly higher than 10^{15} eV the frequency is of only one per m^2 per year. This is reduced to only one per km^2 per year for energies slightly higher than 10^{18} eV, and for energies above 10^{20} eV the frequency of arrival of these particles is estimated to be, at most, of one per km^2 per century. It is understandable, therefore, that the facilities to study ultra-high energy cosmic rays at the Earth's surface must occupy large areas to compensate for their extremely low arrival flow.

The primary cosmic rays can also be observed in situ, as well as the cascades they originate, due to the resulting fluorescence, whose intensity is proportional to the energy of the primary. To get an idea of the extent of this phenomenon, let us bear in mind that a cascade of ultra-energetic cosmic rays contains around 10^{11} particles, that is, one hundred thousand million, which cross the skies at speeds close to the limit c. The fluorescence produced by the cascade is due to the fact that the charged particles, especially electrons and positrons, excite molecules of the atmospheric nitrogen, and these, when de-excited, emit photons in the near ultraviolet range, which are invisible to our eyes, but can be detected with appropriate instruments. Using this technique, in 1991 the High Resolution Fly's Eye Cosmic Ray Detector at Dugway Proving Ground, Utah (USA), observed a cosmic ray with energy of 3.2×10^{20} eV, which holds the record of the energies registered to date,

---

[5]The energy generated in a head-on collision between two particles is simply the sum of the energies of each particle. However, the energy generated in a collision of a particle against a stationary target; for example a cosmic-ray particle against a proton, or neutron, of a molecule in the atmosphere, is given by $\sqrt{m_a^2 c^4 + m_b^2 c^4 + 2 m_a c^2 E_b}$, where m_a is the mass of the stationary particle, m_b the mass of the incident particle and E_b its energy. In the case that E_b is much higher than the energy due to the masses of the particles, a good approximation to this formula is $\sqrt{2 m_a c^2 E_b}$.

although a few more cosmic rays have been observed with similar energies since then.

It should also be noted that in recent times balloon-borne experiments are being carried out with the primary objective of analyzing the Ultra-High Energy Cosmic Rays. We will come back to this below, in Sect. 5.6.1.

5.4   The Pierre Auger Observatory

The Pierre Auger Observatory, located in the Argentinean province of Mendoza, was conceived by Jim Cronin and Alan Watson at the 1991 International Cosmic Ray Conference in Dublin, although its construction only started in the year 2000. By 2003 it had become the largest observatory in the world for the study of cosmic rays, hosting some 500 scientists from 18 countries who participate in the research projects. This observatory combines the two techniques for cosmic ray detection that we have just described and allows the observation of 30 events per year from primaries with energies above 10^{20} eV. The observatory is equipped with an array of 1660 surface detectors (Fig. 5.8)—*Water Cherenkov Detectors* (WCDs)—at a distance of 1.5 km from each other, covering an area of about 3000 km². In 2015 the preliminary design of an upgrade of the observatory to add further capabilities to the surface detectors was issued, and the upgrade was named

Fig. 5.8 Artist's impression of an air shower over a surface Water Cherenkov Detector (WCD) at the Pierre Auger Observatory. The air shower is seen against a real photo showing an aurora. *Credit* A. Chantelauze, S. Staffi and L. Bret

AugerPrime. Shortly afterwards, a small engineering array of 12 upgraded WCD stations built for testing purposes was installed, as seen in Fig. 5.9. The observatory has also a fluorescence detector consisting of 24 telescopes in order to register the arrival of primary cosmic rays entering the Earth's atmosphere.

The observatory's main objective is to study the Ultra-High Energy Cosmic Rays with energies starting at 10^{18} eV, although it can also detect γ rays of similar energies, giving rise to electromagnetic cascades. To be precise, one tries to determine the exact energy of these particles, as well as their nature and place of origin. It happens that there are two crucial problems facing the study of cosmic rays. One is that their origin is unknown, and the other is that the physical mechanisms accelerating particles to such high speeds are also unknown, as noted before. Even so, it is commonly assumed that cosmic rays with energies up to 10^{18} eV could be produced in the remnants of supernova explosions, although some researchers argue that this explanation would only be valid for energies up to 10^{15} eV.

The reason why the sources of the cosmic rays are unknown is that the cosmic-ray particles, being electrically charged, do not travel in a straight line on their way to the Earth. Instead, they follow almost random trajectories under the influence of the magnetic fields present through vast regions of space. Their trajectories also get bent when passing through very intense

Fig. 5.9 Some upgraded WCD stations of the AugerPrime Engineering Array, where one can see the radio antennas mounted on the new Surface Scintillator Detectors, with an active area of 3.8 m², on top of the WCD. *Credit* Nicolás Leal / Pierre Auger Collaboration

gravitational fields, but this is a less relevant effect unless they get trapped, obviously. In short, the direction of arrival of cosmic rays to Earth does not give us any information about where, in which astronomical object or place, the sources that originated them are located.

5.5   The GZK Limit

An important and curious result, in relation to the Ultra-High Energy Cosmic Rays, is the *GZK limit*, also known as *GZK cutoff,* named after the scientists who computed it in 1966: Kenneth Greisen at Cornell University and, independently, Georgi Zatsepin and Vadim Kuzmin from the Lebedev Institute, in Moscow. They analyzed theoretically the interactions between the cosmic ray protons and the photons of the newly discovered Cosmic Microwave Background radiation, CMB. Their calculations predicted that protons with energies greater than 5×10^{19} eV—the GZK limit—interact considerably with the photons of the CMB radiation, generating pions mainly through the process:

$$\gamma_{CMB} + p \rightarrow \pi^0 + p, \tag{5.1}$$

thereby slowing down the protons and reducing their energy drastically (about 20%). This threshold energy implies that there are no cosmic ray protons surpassing it coming to Earth from distances beyond 50 Megaparsecs (about 163 million light years), which is the typical diameter of a supercluster of galaxies. For this reason, the ultra-high energy cosmic ray protons above the GZK limit can only reach the Earth if they come from the supercluster of galaxies to which our Milky Way belongs.

Another interesting aspect of the GKZ limit is that cosmic rays with energies above it, called *Extreme-Energy Cosmic Rays* (EECR), should not be substantially deflected by the action of the surrounding magnetic fields. Therefore, their direction of arrival directly points towards the location where they originated, unlike the case for cosmic rays with energies below the GKZ limit, as discussed above.

5.6   Detectors in Space

Cosmic ray research is also carried out by installing detectors in space, either on board high altitude balloons, attaining between 18 and 40 km in the stratosphere, or by attaching them to satellites or to the International Space

Station. The experiments are equipped with a magnetic spectrometer, whose main function is to separate the detected particles according to their electric charge, and a variety of specialized detectors in order to perform precise measurements. These detectors are, in fact, most appropriate to find and study primary cosmic rays since they are located very high in the atmosphere, or above it, so that essentially all the particles they register consists of primary cosmic rays. Nevertheless, they receive a small amount of secondary radiation from the cascades and their fluorescence, as these spread out in all directions and even upwards.

5.6.1 Balloon-Borne Experiments

As regards the balloon-borne experiments, it is worth mentioning the program BESS (*Balloon-borne Experiment with Superconducting Spectrometer*), a joint project of Japanese and US scientists that was focused on *searching for antimatter* in the cosmic radiation. The researchers of the BESS program were carrying out a systematic measurement of *low energy antiprotons*, hoping to find antihelium nuclei as well, but only antiprotons were detected. The program started in 1993 and had about 10 successful flight campaigns, ending in 2007. The two last flights, called BESS-Polar I and Polar II, which took place in 2004 and 2007, respectively, encircled Antarctica after they were launched from McMurdo Station (Fig. 5.10).

Fig. 5.10 The BESS-Polar I detector shortly before launch in Antarctica, from McMurdo Station, on December 13, 2004. *Credit* Courtesy of NASA and KEK (Japan)

In more recent times, the program SuperTIGER (*Super Trans-Iron Galactic Element Recorder*), from NASA and several US institutions, has flown several times, including a flight over Antarctica between December 2012 and January 2013 that broke ballooning records for duration, staying afloat for 55 days. This project is designed to detect heavy elements in cosmic rays. The last balloon was launched in early December 2019, and on December 31 it had completed its first full revolution—40,000 m above Antarctica.

In the meantime, the program EUSO-SPB (*Extreme Universe Space Observatory–Super Pressure Balloon*), launched its first football stadium-size balloon in April 2017, from Wanaka Airport (New Zealand), in search of air showers of Ultra-High Energy Cosmic Rays (UHECRs). To be precise, searching for the fluorescent tracks of ultraviolet light emitted by the extensive cascades produced by those particles when colliding with atoms in the atmosphere. It is a NASA mission led by Angela Olinto (Fig. 5.11), Professor of Astrophysics at the University of Chicago (USA), in which seven countries participate. The EUSO-SPB experiment will be the first ever to record the fluorescence from the UHECRs by looking down at the atmosphere instead of up, unlike the

Fig. 5.11 Angela Olinto, Professor at the University of Chicago (USA), and Dean of the Division of the Physical Sciences since 1 July 2018. She is the Principal Investigator of current NASA sub-orbital and space missions to discover the origins of the Ultra-High Energy Cosmic Rays and neutrinos. In particular, the EUSO-SPB program and the future POEMMA mission. *Credit* John Zich. Courtesy of Angela Olinto and University of Chicago

ground-based experiments. This broadens the field of observation, improving the detection rate of these rare "visitors".

This balloon flight was cut short, however, after 12 days, because there was a leak. As a consequence, it began undergoing substantial drops in altitude at night, regaining its designed altitude of about 33.2 km during the day as the temperatures increased. In any case, the EUSO-SPB1 experiment performed well, and more than 60 GB of data could be downloaded. Although the data set revealed different types of events, no obvious tracks of UHECRs, that is, of cosmic rays with energies starting at 10^{18} eV, were found. The next EUSO-SPB2 experiment is planned to be launched in 2022, and is expected to make several trips around the Antarctic region over 100 or more days.

5.6.2 Experiments on Board Satellites

The pioneers in cosmic ray research from space satellites were the Soviets. They sent the *Proton* series, a total of four satellites which were operating between 1965 and 1969, with the main objective of studying high energy cosmic rays. The detectors and other devices on board were used to identify them and to analyze the intensity and energy spectrum of the cosmic electrons up to energies of 10^{14} eV, as well as to measure the intensity and energy of the γ rays over 5×10^7 eV. The first satellite of the series was Proton 1, which was launched on July 16, 1965, and reentered the atmosphere less than three months later, on October 11. The Proton 2 and 3 satellites had similar short lifetimes, but Proton 4 remained in orbit for 250 days.

PAMELA
The PAMELA (*Payload for Antimatter Matter Exploration and Light-nuclei Astrophysics*) experiment, in which several European institutions were involved, was the first experiment installed on a satellite to investigate the primary cosmic rays *with special emphasis on its antimatter component*. This component seems to exist only in the form of positrons e^+ and antiprotons $\overline{p}$, as no other type of antimatter has been detected so far. This situation might change in the next few years, however, since the AMS-02 experiment has found some candidates for antihelium nuclei (see Sect. 5.6.3).

The PAMELA experiment was installed on the Russian Resurs-DK1 satellite (Fig. 5.12), which was launched into space in June 2006 aboard a Soyuz rocket from the Baikonur Cosmodrome of the former Soviet Union. It was placed in an elliptical orbit at an altitude between 350 and 610 km, although since September 2010 it was orbiting at 580 km until the end of its mission in 2016. The detector, with a height of 1.30 m and a weight of 470 kg, looked like a dice cup, as can be seen in Fig. 5.12. Other objectives of this

Fig. 5.12 Artist's illustration of the PAMELA detector, that looks like a dice cup, on board the Russian Resurs-DK1 satellite. It was launched on 15 June 2006 on a Soyuz rocket from the Baikonur Cosmodrome of the former Soviet Union, and finished its mission in 2016. *Credit* Courtesy of Piergiorgio Picozza, on behalf of the PAMELA collaboration

project were to analyze the effects of the solar activity on the propagation of cosmic rays and to investigate the inner Van Allen belt in the Earth's magnetosphere—the magnetic field surrounding the Earth.

In the ten years of operation, the PAMELA experiment obtained important results in determining the composition of the cosmic rays, as well as in the study of their energies. In addition, it successfully analyzed the effects of the solar activity—the heliosphere[6]—on the propagation of cosmic rays, which had never been carried out before. But the most impressive results were harvested in relation to antimatter, even though only positrons and antiprotons were found. This was already a relevant piece of information, nevertheless, since it allowed to improve the estimates on the antihelium abundance with respect to helium up to 10^{-7}. This means that for every antihelium atom that exists in the local Universe there are no less than 10,000,000 helium atoms.

The first result, obtained already in 2008, was that the data showed an excess of positrons, compared to previous estimates, which permeate the space in the range of energies between 10 and 60 GeV. This finding was totally unexpected but was corroborated shortly afterwards by the Fermi Gamma-Ray Space Telescope, placed in orbit in June 2008. According to many

---

[6]The heliosphere is a bubble-like region created by the Sun that surrounds the solar system. It consists of the solar magnetic field and the plasma (mostly protons, electrons and α particles) emitted by the Sun, called the "solar wind".

physicists, the observed positron surplus could be due to the annihilation of dark matter particles with each other, a possibility that gave rise to hundreds of scientific articles, although some other explanations were proposed as well within the framework of conventional Physics and Astrophysics.

A second result was that, in 2011, the PAMELA collaboration confirmed the suspicion that the inner Van Allen radiation belt, the closest to the Earth, might harbor a significant amount of confined antiprotons. These would be produced by energetic cosmic rays colliding with molecules in the upper layers of the atmosphere since the antiprotons thus produced can bounce upwards and get caught in the Earth's magnetic field. Indeed, the PAMELA experiment did find these antiprotons as part of the inner belt and also concluded that their flux exceeded that of the cosmic ray antiprotons by a factor of 1000. The inner Van Allen belt is, therefore, the most abundant source of antiprotons near the Earth (Fig. 5.13).

POEMMA

The POEMMA (*Probe Of Extreme Multi-Messenger Astrophysics*) experiment is a planned NASA mission, which will put two satellites in orbit in order

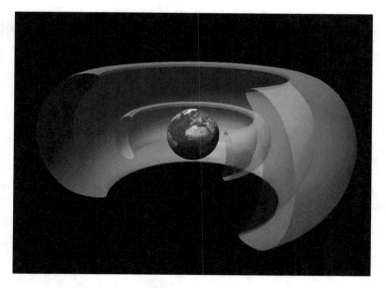

Fig. 5.13 The radiation belts, discovered in 1958 by James Van Allen, are two broad regions of very intense radiation in a toroidal shape around the Earth, where particles trapped by the Earth's magnetic field, mainly protons in the inner belt and electrons in the outer belt, are moving at high speeds spiraling between the planet's magnetic poles. In 2011 the PAMELA experiment found a band of antiprotons on the inner belt whose flux exceeded that of the cosmic ray antiprotons by a factor of 1000. *Credit* Courtesy of NASA

to study Ultra-High Energy Cosmic Rays with more advanced telescopes than those of the balloon EUSO-SPB program. The team of scientists and NASA engineers, led also by Angela Olinto (Fig. 5.11), is already designing this mission with high expectations about its capabilities for discovery.

5.6.3 The AMS-02 Experiment

At present, the most powerful particle detector operating in outer space is the *Alpha Magnetic Spectrometer* (AMS-02), which is installed in an external module of the ISS (International Space Station), as can be seen in Fig. 5.14. Its main purpose is to *search for antimatter*, as well as for signals of dark matter annihilation, although it also makes various types of measurements on cosmic rays with unprecedented precision.

The AMS-02, which weighs 7500 kg, was assembled at CERN, where it passed its final tests in August 2010. Then it was transported to the Kennedy Space Center at Cape Canaveral (Florida, USA) and from there it was delivered to the ISS on board the Space Shuttle Endeavor, on its last flight on May 16, 2011 (Fig. 5.15). Three days later it was already sending data to Earth, which were received by NASA in Houston and then relayed to the AMS Payload Operations Control Center (POCC), at CERN, for analysis.

Fig. 5.14 The Alpha Magnetic Spectrometer (AMS-02) on board the International Space Station (ISS). Its primary purpose is to search for antimatter, as well as for signals of dark matter annihilation, although it also makes various types of measurements on cosmic rays with unprecedented precision. *Credit* Courtesy of NASA

Fig. 5.15 Launch of the AMS-02 detector from the Kennedy Space Center on 16 May 2011, marking the 25th and final voyage of the Space Shuttle Endeavour to the ISS. *Credit* Courtesy of NASA

The AMS-02 experiment is carried out by an international collaboration of 56 institutions from 16 countries, with the participation of around 600 scientists. The leader of the experiment is Samuel Ting (Fig. 5.16), from CERN and the Massachusetts Institute of Technology (MIT), USA, who proposed this project in 1995, just after the creation of antihydrogen atoms at CERN. It must be said that without his dedication and determination this experiment would not have been carried out, since the project suffered a whole series of mishaps and unpleasant surprises. It must also be said that Ting surrounded himself with excellent collaborators who embraced this undertaking.

Following Ting's proposal, a test prototype named AMS-01 was built, which was launched into space in 1998 and flew aboard the Space Shuttle Discovery for 10 days, including a four day docking phase to the Soviet MIR Space Station. This prototype was successful, which led to the approval of the construction of the AMS-02 detector, whose core is a gigantic 1200 kg magnet. This magnet is surrounded by several types of detectors, specialized in various types of measurements, which register around 1000 cosmic rays per second. In fact, on 8 May 2017, the total number of recorded events reached 100 thousand million (10^{11}) and while this book was being finished (November 2020) more than 168 thousand million cosmic ray events had been collected. The AMS-02 detector has been recently upgraded to be able to operate during the entire lifetime of the ISS, which is expected to be at least until 2030.

Fig. 5.16 Samuel Chao Chung Ting at the Kennedy Space Center with the Space Shuttle Endeavour on his back, in 2011. Professor of Physics at MIT (USA) and researcher at CERN, he has led several international experiments in Particle Physics in accelerator laboratories, and is the Principal Investigator of the AMS-02 experiment installed in the ISS. He received the 1976 Nobel Prize in Physics when at Brookhaven National Laboratory (USA), shared with Burton Richter from SLAC (USA), for the discovery of the subatomic particle charmonium, consisting of a bound state of a charm quark c and its antiquark. He also participated in the discovery of the antideuteron at Brookhaven (1965), and of the gluons at DESY (Germany, 1979). *Credit* Courtesy of the AMS collaboration

As for the results accomplished by the AMS-02 experiment in the nine years of operation, it should be noted first that the very precise analysis of the composition, energy and flow of the cosmic rays is in itself a very important achievement. Not only because it provides us with scientific data about the production of cosmic rays and the mechanisms of their acceleration and propagation, but also for purely practical reasons. For example, a better knowledge of the cosmic rays in space can be used to improve the safety of astronauts and satellites, since cosmic radiation is a major obstacle for space travel in general, and for sending human beings to Mars in particular. In what follows, we will say a few words about the most relevant findings obtained by the AMS collaboration regarding antimatter.

Positrons e^+
The AMS-02 experiment has confirmed the excess of high energy positrons found by the PAMELA collaboration, and the AMS researchers also think

that such excess might be a sign of dark matter annihilation, among other possibilities. Let us have a closer look at this.

In the article *Towards Understanding the Origin of Cosmic-Ray Positrons*, published in 2019 in "Physics Review Letters", the AMS collaboration presented the precision measurements based on 1.9 million cosmic ray positrons registered by the AMS-02 detector from May 2011 to November 2017. The researchers reported that, even though there is an important production of secondary positrons in collisions of cosmic rays with interstellar gas, the data show that this production is predominant only for low energy positrons and therefore other mechanisms must exist in order to account for the excess of high energy positrons. Moreover, they proposed that this positron surplus must originate, either as by-products of dark matter annihilation, or from astrophysical sources where they are accelerated, such as pulsars.

This proposal is based on the peculiar dependence of the flux of positrons with respect to their energy (Fig. 5.17), that suggests the existence of a source of high energy positrons with a well defined energy range; in particular with a cut-off where the positron production stops. As can be seen in the plot, the positron flux recorded by the AMS-02 detector shows a significant excess starting from about 5 GeV compared to the predictions of two models of positron production in collisions of cosmic rays with interstellar gas. Then

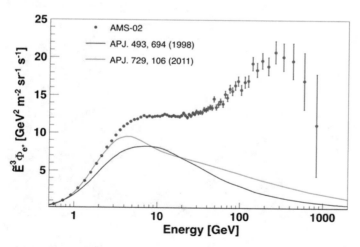

Fig. 5.17 Dependence on the energy of the flux of positrons observed by the AMS-02 experiment (red dots) versus the same dependence predicted by two models of positron production in collisions of cosmic rays with interstellar gas. The positron flux registered by the AMS-02 detector shows a significant excess compared to these models, as was already reported by the PAMELA experiment. *Credit* Courtesy of the AMS collaboration

the flux climbs to 284 GeV and starts dropping sharply until it disappears at about 810 GeV. This pattern shows that not only the positron excess by itself, but also the sharp drop-off and the cutoff at 810 GeV, all are inconsistent with the models of positron production in collisions of cosmic rays, which exhibit much simpler patterns. Certainly, this result calls for a source of high energy positrons, as displayed in Fig. 5.18.

Another piece of information about possible sources of the high energy positrons comes from the recent article *Detection of a γ-ray halo around Geminga with the Fermi-LAT data and implications for the positron flux,* published in "Physics Review D" in December 2019. This article was already discussed in some detail in Sect. 4.6 of the previous chapter, and its major conclusions can be summarized in just one phrase: The excess of positrons observed near Earth is coming from pulsars in the Milky Way, and the pulsar Geminga alone could be responsible for as much as 20% of the high energy positrons detected by the AMS-02 experiment.

Antiprotons $\overline{p}$

In the article *Antiproton Flux, Antiproton-to-Proton Flux Ratio, and Properties of Elementary Particle Fluxes in Primary Cosmic Rays Measured with the Alpha Magnetic Spectrometer on the International Space Station*, published in

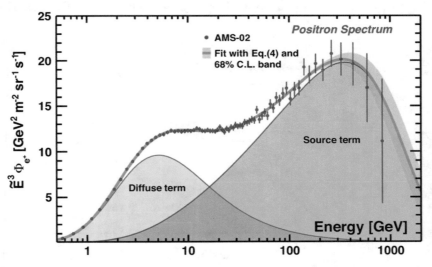

Fig. 5.18 In the entire energy range, the positron flux is well described by the sum of a "Diffuse term" associated with the positrons produced in collision of cosmic rays, which dominates at low energies, and a new "Source term" of positrons, which dominates at high energies. The latter could consist of by-products of dark matter annihilation or of various astrophysical objects such as pulsars. *Credit* Courtesy of the AMS collaboration

2016 in "Physics Review Letters", the AMS collaboration presented precision measurements of the antiproton flux and the antiproton-to-proton flux ratio in primary cosmic rays based on 3.49×10^5 antiproton events and 2.42×10^9 proton events. The dependence of these fluxes on the energy is usually expressed as the dependence on the *rigidity* R instead, which is the energy per unit of electric charge: $R = E/q$. Accordingly, the measurements of the fluxes were shown as a function of the absolute value of R, ranging from 1 to 450 GV. The results showed that, for R between 60.3 GV and 450 GV, the antiproton-to-proton flux ratio has a mean value of 1.81×10^{-4}, and the antiproton-to-positron flux ratio has a mean value of 0.48. In both cases, the flux ratios were almost constant, i.e. R independent, in such interval.

It is also worth mentioning that, since 2015, various groups have identified an excess of antiprotons in the fluxes of cosmic rays observed by the PAMELA and AMS experiments. Moreover, they propose that this antiproton excess could originate in the annihilation by-products of dark matter. One recent article with such proposal is *A Robust Excess in the Cosmic-Ray Antiproton Spectrum: Implications for Annihilating Dark Matter,* by Dan Hooper from Fermilab and collaborators, published in 2019 in "Physics Review D". After an exhaustive analysis of the AMS data, the researchers confirm the presence of an antiproton excess consistent with that arising from the annihilation of dark matter particles with masses between 64 and 88 GeV. They also point out that the same range of dark matter models that are favored to explain the antiproton excess can also accommodate an excess of γ rays observed from the Galactic Center.

The AMS collaboration, in turn, will publish soon a review in "Physics Reports" where, among many other results, some comparisons will be presented between the antiproton flux and various predictions from some dark matter models.

Antihelium nuclei

In December 2016, Samuel Ting presented at CERN the results of the first five years of the AMS-02 experiment. To the general surprise, he announced that they had found about five particles, one a year or so, that are candidates for antihelium nuclei of the isotope $^3\overline{\text{He}}$. Since antimatter particles have the same mass and opposite charge as their matter counterparts, this means that the AMS-02 detector could have recorded particles with electric charge equal to -2 and with a mass very similar to that of the helium isotope ^3He, with two protons and one neutron. Less than two years later, in April 2018, the AMS-02 collaboration had about six candidates for $^3\overline{\text{He}}$ nuclei as well as two for $^4\overline{\text{He}}$ nuclei—two possible anti-alpha $\overline{\alpha}$ particles.

These results were presented by a member of the AMS collaboration—Vitaly Choutko from MIT (USA)—in the contribution *AMS Heavy Antimatter* at the meeting "AMS days at La Palma", in La Palma (Canary Islands, Spain).

However, there is always the possibility that these signals are due to background noise, because of their scarcity. Hence many more data are needed to be able to decide whether those candidates are really antihelium nuclei or background, and this can take several years still. For this reason, the AMS collaboration has not yet published any figures concerning the antihelium candidates. In any case, these results have allowed the researchers to improve the estimates on antihelium abundance with respect to helium by one order of magnitude, so that it is now 10^{-8}. That is, the rate is one antihelium to 100 million helium.

The discovery of $^3\overline{\text{He}}$ and/or $^4\overline{\text{He}}$ nuclei in cosmic rays would create a great expectation, even though these nuclei have already been artificially created in our labs, as explained in Chap. 4. The reason for this lies in the possibility that these antinuclei originated from nucleosynthesis of primordial antimatter, created in the first moments of the Universe's existence, or from "antistellar" nucleosynthesis inside antimatter stars. Nevertheless, it cannot be excluded that extremely energetic known astrophysical processes, such as supernova explosions, neutron star mergers, quasars and other active galactic nuclei (AGN), among others, may produce antihelium nuclei.

Accordingly, we could only be sure that primordial antimatter exists if antinuclei heavier than those of antihelium were found, because they certainly cannot be created in any astrophysical processes taking place in astronomical objects made up of matter. In fact, the AMS researchers would like to make sure they can also detect anticarbon and antioxygen nuclei before publishing any results about the antihelium candidates. The detection of those heavier antinuclei would be most striking, certainly, since they could only have been created in the nuclear furnaces inside antistars. Consequently, the observation of anticarbon or antioxygen nuclei in nature would provide a definitive proof of the existence of antimatter stars and therefore of the survival of primordial antimatter at the beginning of the Universe.

However, the view that extremely energetic known astrophysical processes may produce antihelium nuclei is being questioned by more and more scientists, especially by those acquainted with the details of antihelium production in the laboratory. In brief, they think that the difficulties associated to antihelium production are such that it must be essentially impossible for nature to create antihelium nuclei out of very energetic processes in astrophysical objects made of matter. In this vein the article *Where do the AMS-02 antihelium events come from?* was written, by Joseph Silk and collaborators, and was published in "Physics Review D" in 2019. They argue that usual astrophysical explanations, as well as dark matter annihilations, face real difficulties in order to explain the possible antihelium nuclei reported by the AMS collaboration. For this reason, they consider the possibility that these antinuclei would originate from large antimatter dominated regions in the form of antigas clouds or antistars. We will return to this topic in Chap. 7, Sect. 7.1.3, where the possibility of existence of antimatter structures such as antistars and antigalaxies will be discussed.

6

Particle Physics Accelerators

6.1　Generalities

Particle accelerators are instruments designed to accelerate corpuscles endowed with electric charge. These corpuscles can be subatomic particles, both elementary or hadrons, as well as ions and atomic nuclei. In order to accelerate them, they are subjected to very intense electric fields until they reach speeds very close to c (in the most powerful accelerators, they are practically indistinguishable from c). The accelerators also come equipped with intense magnetic fields, which serve both to confine the corpuscles in very thin beams and to curve their trajectories, in the case of circular accelerators.

But not all subatomic particles can be accelerated, because for this to be possible they not only must have electric charge, but they must also live long enough for it to make sense to try to accelerate them. Hence, we have that neutrons and neutrinos cannot be accelerated, despite being very long-lived, as they lack electric charge; while pions, kaons, and the W^+ and W^- bosons, among many other particles, cannot be accelerated either, despite having electric charge, due to the fleeting nature of their existence.

Depending on how the accelerated corpuscles will be used, greater or lesser energies are required. For example, a beam can be aimed at a metal target to produce certain particles such as positrons, antiprotons, or muons, but it can also be used to bombard atomic nuclei in Nuclear Physics experiments; or to irradiate tumors in oncological therapies, as well as in a variety of other techniques.

© Springer Nature Switzerland AG 2021
B. Gato-Rivera, *Antimatter*,
https://doi.org/10.1007/978-3-030-67791-6_6

6.2   High Energy Collisions

Although the collisions of a particle beam against a stationary target have been very successful, and are still broadly used, in order to reach the highest energies two beams of particles traveling in opposite directions are needed, which are made to collide head-on. These collisions take place at specific points of the accelerator, where detectors are installed in order to record the resulting products.

To create matter and antimatter particles in the accelerators, different from those being accelerated, the energy reached in the collisions must exceed the energy corresponding to the mass of the particle to be created, applying the formula $E = m c^2$. As a matter of fact, it is often necessary to surpass more than twice this amount, in the processes in which a particle is created together with its antiparticle, and even much more than twice if pairs of hadrons are created, like a proton-antiproton pair, $p\,\bar{p}$, for very technical reasons which we will not discuss here. The particles[1] that collide usually disappear and other more massive ones are created among the by-products, in which *particles and antiparticles appear in similar numbers*. The heavy particles are unstable and decay very quickly, giving rise to other particles that are lighter. These can interact with each other and, in case of being unstable, they also decay until, finally, stable or semi-stable particles are obtained which are registered by the detectors. In addition, if the unstable particles that were created live long enough, such as pions or muons, the detectors can register them too, together with the products of their decays.

A crucial aspect of the processes between particles is that there are quantities that have to be conserved at each step, or intermediate stage, as well as in the final configuration. These quantities are the electric charge, the energy and the linear momentum (the mass multiplied by the velocity), among others.

However, it turns out that in processes between particles there is a property that must be satisfied only by the initial particles and the particles in the final configuration, but not by the particles present at the intermediate stages. The particles in these intermediate configurations are called *virtual*, as opposed to the real particles in the initial and final configurations, because they cannot be detected by any device or apparatus. The property referred to is a relationship involving the mass, energy, and speed of particles—the classical equations of

---

[1] Observe that, depending on the context, the term "particles" refers to particles and antiparticles alike, as "antiparticle" is simply a short name for "antimatter particle". Nevertheless, the term antiparticle is more general, since mesons and gauge bosons, which are neither matter or antimatter, can be viewed as particles and antiparticles too.

motion—and when satisfied, the particle is said to be on the mass shell, or simply *on shell*. Thus, saying that a particle is real and can be registered by a detector is equivalent to saying that it is on shell.

Virtual particles, on the contrary, do not satisfy the classical equations of motion and are said to be off the mass shell, or *off shell*. This renders the intermediate stage configurations less restricted than the final configurations of the real particles and implies that the virtual particle masses are not the same as those of real particles; in fact, it does not even make sense to assign any mass to virtual particles at all. Even so, virtual particles and their quantum fluctuations contribute to all processes between particles as well as to the vacuum energy.

6.2.1 Electron–Positron Collisions

As was explained in Chap. 2, in e^+e^- collisions at high energies not only photons are created, as happens at low energies. Not only that, but even in the case where initially the positron e^+ and the electron e^- annihilate each other producing a virtual photon, this photon disappears immediately because its high energy is invested in exciting other quantum fields and thus create particles with mass; for example, a pair of W^+ and W^- bosons, as can be seen in the Feynman diagram on the left of Fig. 6.1. Now, due to electric charge conservation, the e^+e^- collisions, with total electric charge equal to zero, can only result in intermediate stages and final products which are electrically neutral; for example, photons and particle-antiparticle pairs, like the bosons W^+ and W^- of our example, although they could also be quark-antiquark pairs, or again a pair e^+e^-, among other possibilities.

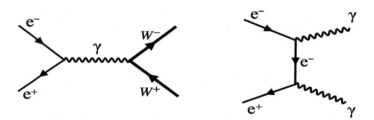

Fig. 6.1 Feynman diagrams corresponding to electron–positron annihilation, where the time goes from left to right. In the diagram on the left, the energy of the virtual photon is invested in creating a pair of W bosons, which decay producing lighter particles. But this energy could also have been invested in creating a quark-antiquark pair or another pair of electric charged particles, including another electron–positron pair

And there is more, because in the event that the e^+e^- annihilation gives only photons as products, the conditions of energy and linear momentum conservation require more than one photon in the final configuration. Consequently, the e^+e^- annihilation cannot give a single photon as final result, but there must be at least two photons, as can be seen in the Feynman diagram on the right of Fig. 6.1 and in Fig. 2.2 in Chap. 2.

6.2.2   Hadron Collisions

Collisions between hadrons, like $p\,p$, $p\,n$, or $p\,\overline{p}$, are of great complexity because, although hadrons are described as being composed of two or three quarks or antiquarks, they actually harbor a whole multitude of components. For example, if we inspect a proton at very short distances, what we find is that it consists of three *valence* quarks and a sea of quarks and antiquarks of the three families exchanging gluons. The latter, in turn, are continuously generating quark-antiquark pairs and more gluons. One says that a proton consists of two quarks u and one quark d because in the total count of quarks and antiquarks only these three quarks are unpaired. The unpaired quarks or antiquarks within the hadrons are called valence quarks or antiquarks and are the most involved ones in interactions with other particles, along with gluons.

Hence, the collisions between hadrons are translated into collisions between quarks, antiquarks and gluons; that is, in collisions of the types: $q\,q$, $q\,\overline{q}$, $\overline{q}\,\overline{q}$, $q\,g$, $\overline{q}\,g$, and $g\,g$. Now, by means of the so-called *hadronization*, quarks, antiquarks and gluons always organize themselves into hadrons, and this is due to the confinement of the color charge, as explained in Chap. 2, Sect. 2.5.3. As a consequence, even though the initial hadrons disappear at the very beginning in high energy collisions, the same type of hadrons often reappear in the final products. For instance, one possible process for obtaining antiprotons by colliding a beam of protons, of at least 6 GeV, against a metal target (copper, nickel, or iridium), is:

$$p + p \rightarrow p + \overline{p} + p + p. \tag{6.1}$$

This process can be viewed, so to speak, as a proton in the beam colliding with a proton of an atomic nucleus, the initial protons "reappearing" while the energy of the collision is invested in creating a proton-antiproton pair. Similarly, in collisions of high energy protons in cosmic rays against atomic nuclei of the atmosphere, antiprotons are created through the analogous process: $p + A \rightarrow p + \overline{p} + p + A$, where A represents the atomic nucleus, as shown in Chap. 4, Eq. (4.3).

6.3   Brief History of Particle Physics Accelerators

The Cyclotron

The first accelerator to be built was the cyclotron of Ernest Lawrence (Fig. 6.2), in 1931, at the University of California at Berkeley. It was a device of very small size, about 10 cm in diameter, which was followed by several more in the same and the following decade. Each cyclotron was larger and more powerful than the previous, the last one reaching 4.67 m in diameter. In a cyclotron, the corpuscles move in a circular fashion spiraling from the center outwards because these devices are equipped with an electromagnet which makes the trajectories curve in this way. It so happened that the mechanism for operating these accelerators had already been proposed between 1926 and

Fig. 6.2 Ernest Lawrence (1901–1958) when he received the 1939 Nobel Prize in Physics for his invention of the cyclotron. *Credit* The Nobel Foundation

1929 by several scientists in Germany, but for various reasons its construction was not carried out, although it was even patented by Leo Szilard in 1929 while working at the Humboldt University in Berlin.

The purpose of these instruments was for research in Nuclear Physics and the corpuscles used were ionized atoms. In the most powerful cyclotron, built in 1939, plutonium, neptunium and other more transuranic elements were discovered; in fact, many of their isotopes were synthesized. In addition, during the 1940s, cyclotrons began to be used to separate radioactive isotopes, as in the Manhattan project that led to the first atomic bombs. Yet the energy reached in those accelerators did not exceed several hundred MeV, which means that the velocities of the accelerated corpuscles were not relativistic, i.e. they were well below the speed limit c. This was not a disadvantage in relation to research in Nuclear Physics, nor with respect to their use in proton radiation therapy, but Particle Physics, for its development and research, needed much higher energies than cyclotrons could provide. This led to the birth of the next generation of accelerators in the 1950s: the *synchrotrons*, in which particles are accelerated in a ring of constant radius.

In April 1948, the US Atomic Energy Commission approved the construction of two large particle accelerators of synchrotron type—two proton synchrotrons—one at the Brookhaven National Laboratory (BNL), in New York, named Cosmotron, and the other at the Lawrence Berkeley National Laboratory (LBNL), in California, named Bevatron. Let us say a few words about these and several other large accelerators which were subsequently built all over the world.

BNL: The Cosmotron

The Cosmotron at Brookhaven (Fig. 6.3), built in 1952, was the first accelerator to reach energies in the GeV (10^9 eV) range, accelerating protons up to 3.3 GeV. The ring was 72 m long and counted with 288 magnets weighing 6 tons each. Its main purpose was to produce all the mesons that had only be seen previously in cosmic ray cascades (essentially, pions and kaons). In fact, the name Cosmotron comes from "Cosmic cyclotron" and it was the first accelerator that was able to produce all positive and negative mesons known to exist in the secondary cosmic radiation. The Cosmotron ceased operation in 1966 and was dismantled in 1969.

BNL: The AGS and the RHIC

Brookhaven's next large accelerator was the Alternating Gradient Synchrotron (AGS), which became the most powerful proton synchrotron in the world between 1960 and 1968, reaching 33 GeV of energy. The muon neutrino and the charm quark and its antiquark were discovered in this accelerator, in 1962 and 1974 respectively, which is still in operation and used as injector

Fig. 6.3 The Cosmotron at Brookhaven National Laboratory (BNL) in Long Island, New York. It was a proton synchrotron 72 m long, the first large accelerator ever built. It was operating from 1952 until 1966. *Credit* Courtesy of Brookhaven National Laboratory

for the most important accelerator today at this site, the Relativistic Heavy Ion Collider (RHIC). As explained in Chap. 4, this machine accelerates heavy ions, mainly gold ions, and makes them collide head-on with each other. Its ring is 3.86 km long, and its primary purpose is to reproduce and study the conditions in the first instants of the Universe's existence, which are referred to as quark-gluon plasma. RHIC started operating in 2000, and in 2010 was able to synthesize antialpha particles $\bar{\alpha}$ for the first time.

LBNL: The Bevatron

The Bevatron at Berkeley (Fig. 6.4) was constructed two years after the Cosmotron. Beginning in 1954 it was operating for almost 40 years, until 1993. It was about 130 m long and reached almost twice the energies of the Cosmotron since its major objective was to create antiprotons in order to confirm Dirac's prediction of 1931. This required a proton beam of at least 6.2 GeV and, indeed, the antiproton $\bar{p}$ was produced in this accelerator in 1955, the antineutron $\bar{n}$ following in 1956, as we saw in Chap. 4. The Bevatron should also be remembered as the accelerator where the study of relativistic nuclear collisions began, being therefore the precursor of RHIC, JINR (see below) and CERN, in this respect.

The construction of the large synchrotrons in the USA started the race for the discoveries of subnuclear particles and, as a consequence, a wide variety of large accelerators were developed all over the world in the 1950s and 1960s for research in Particle Physics and Nuclear Physics. The construction of these

Fig. 6.4 Two physicists, Edwin McMillan and Edward Lofgren, standing on the shielding of the Bevatron. *Credit* Courtesy of Lawrence Berkeley National Laboratory

accelerators nucleated the work of many physicists, both theorists and experimentalists, and gave rise to several sizeable national and international research centers, with collaborations involving many countries.

CERN: The Synchrotrons PS and SPS

In the first place, in 1954 the European Organization for Nuclear Research (CERN) was created in Switzerland, near Geneva, by a convention of 12 European countries. Its first large accelerator was the Proton Synchrotron (PS), 628 m long, which in 1959 was able to reach 28 GeV of energy. It is still in use as pre-accelerator of the LHC (Fig. 6.5) and for other purposes as well, including the production of antiprotons to create atoms of antihydrogen in the Antimatter Factory. CERN's second and much larger accelerator was the Super Proton Synchrotron (SPS), 6.9 km long, installed deep underground, that reaches a maximum energy of 450 GeV per beam. It came into operation in 1976 and is still at work as pre-accelerator supplying protons and lead nuclei to the LHC (Fig. 6.5). From 1981 to 1984 the SPS was used as proton-antiproton ($p\,\bar{p}$) collider, the first one in the world, achieving in 1983 the discovery of the $W^{\pm}$ and Z^0 bosons.

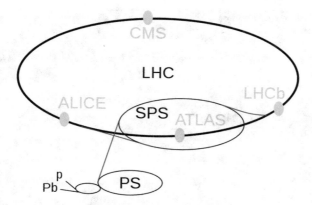

Fig. 6.5 Sketch of the CERN accelerators in the acceleration chain ending at the LHC, with the four main detectors along the LHC ring: ATLAS, CMS, LHCb and ALICE. The beams of protons (p) or ions (lead nuclei Pb) start their journey at two small linear accelerators, which bring them to the PS Booster (the small, unmarked circle). From there they are subsequently injected and accelerated, first in the Proton Synchrotron (PS), then in the Super Proton Synchrotron (SPS), and finally in the LHC ring. *Credit* Courtesy of Arpad Horvath and Wikimedia Commons (Creat. Comm. 2.5 license)

CERN: The Large Colliders LEP and LHC

The next, and gigantic, accelerator at CERN was the Large Electron Positron Collider (LEP), also of synchrotron type, 26.7 km long, which was installed in an underground tunnel 100 m deep. As indicated by its name, LEP was an electron–positron ($e^- e^+$) collider. It started functioning in 1989, reaching a maximum total energy of 209 GeV a few years later. Its major achievement was to prove that there are only three light neutrinos and therefore only three families of fermions. It was shut down in November 2000, to be dismantled and to proceed with the construction of the Large Hadron Collider (LHC) in the same tunnel (Fig. 6.6). The LHC is the largest and most powerful accelerator in the world. After an accident in September 2008 that resulted in a whole year of repairs, the LHC became operational in November 2009. About two years later it produced the Higgs boson, discovery that was announced in July 2012. In its second stage of functioning, it reached a maximum energy of 13 TeV (13,000 GeV) when acting as a proton-proton (p p) collider, although it also accelerated lead nuclei at a maximum energy of 8.16 TeV. We will return to these accelerators in the next section, where we will give an overview of CERN.

JINR: The Synchrophasotron and the Nuclotron

The creation of CERN in 1954 was followed shortly after by the creation in 1956 of the Joint Institute for Nuclear Research (JINR) in Dubna, 120 km

Fig. 6.6 Section of the Large Hadron Collider (LHC) at CERN in the 26.7 km tunnel, 100 m deep, across the border between France and Switzerland. *Credit* Courtesy of CERN

north of Moscow, which became a common project of 12 Socialist countries. In 1957, the JINR constructed the Synchrophasotron, a synchrotron built entirely out of Soviet technology, that operated from 1957 until 2003. Its final energy accelerating protons, and later deuterium nuclei, was 10 GeV. Research in the Laboratory of High Energies (LHE) is currently carried out at JINR with another synchrotron, the Nuclotron, which is a superconducting accelerator of nuclei, built in 1992. It should also be underlined that many of the transuranic elements discovered in the world were synthesized at JINR, which is why the International Union of Theoretical and Applied Chemistry named the 105th element *Dubnium*. The synthesis of the 119th and 120th elements of the periodic table is still underway, and a new laboratory has been built at JINR to achieve this task - the Factory of Superheavy Elements - which makes use of a cyclotron. Curiously, its inauguration, in 2019, coincided with the "International Year of the Periodic Table of Chemical Elements".

DESY: The Synchrotrons DESY, PETRA and HERA

Again in Europe, in 1960 began the construction of the first particle accelerator in Germany. It was the Deutsches Elektronen Synchrotron (DESY), located in Hamburg. On January 1, 1964 the first electrons were circulating in the synchrotron, reaching 7.4 GeV. Later, DESY gave its name to the entire research center, where several other accelerators were constructed over the years. In 1978, the Positron–Electron Tandem Ring Accelerator (PETRA) started running as an electron–positron (e^-e^+) collider. Very soon, in the summer of 1979 one of the PETRA's detectors discovered the gluons—the mediators of the strong interactions. Also worth mentioning is the Hadron Electron Ring Facility (HERA), that started operating in 1992. It was the largest synchrotron at DESY, with a length of 6336 m, and eleven countries participated in its construction in addition to Germany. HERA was the only accelerator in the world that was able to collide protons with either electrons or positrons, which is why it was possible to study the structure of protons and the properties of quarks up to 30 times more accurately than before. It was decommissioned in 2007. DESY now performs scientific research in three major areas: Particle Physics, Photonics, and the construction and development of particle accelerators.

SLAC: The linear accelerator SLAC Linac

Although most of the large accelerators were circular, of synchrotron type, there were also some linear ones, among which we should single out the Stanford Linear Accelerator, known as SLAC Linac, with a length of 3.2 km. Constructed in 1966, it was the longest linear accelerator that has ever existed (Fig. 6.7). It accelerated electrons and positrons, reaching maximum energies of about 50 GeV. It was located at the Stanford Linear Accelerator Center (SLAC), in California, whose name changed in 2008 becoming the SLAC National Accelerator Laboratory. In 1968, by colliding electron beams from the Linac against a fixed target, in order to investigate the internal structure of protons and neutrons, the first evidence of the existence of quarks was obtained; in particular, the up, down and strange quarks were discovered. And in 1974 a bound state of the charm quark and its antiquark was discovered (by a team in SLAC and another independent team in Brookhaven).

SLAC: The accelerators SPEAR and SLC

Moreover, in 1976 the tau particle of the third family of fermions was created at the Stanford Positron Electron Asymmetric Rings (SPEAR), which was

Fig. 6.7 Aerial view of the SLAC National Accelerator Laboratory, showing the 3.2 km building housing the beamline of the SLAC Linac accelerator. *Credit* Courtesy of Peter Kaminski and Wikimedia Commons

built in 1972. It consisted of a ring about 250 m long, in which beams of electrons and positrons collided head-on with energies of 4 GeV. Finally, in 1989 extensive modifications in the original Linac resulted in the Stanford Linear Collider (SLC), colliding electrons with positrons with a total energy of about 100 GeV. SLAC also hosted the experiment BABAR in order to investigate the decay rate of the B mesons, as we will discuss in next chapter. Afterwards, SLAC underwent a shift from a Particle Physics laboratory to an interdisciplinary research center.

Fermilab: The Main Accelerator

In 1969 another High Energy Physics laboratory was founded in the USA. It was the National Accelerator Laboratory, in Batavia, some 60 km from Chicago, which became worldwide known as Fermilab (Fig. 6.8) after the name Fermi was added in 1974 to honor the physicist Enrico Fermi (Fig. 6.9). It was equipped with the highest energy proton synchrotron in the world, called the Main Accelerator, 6.3 km long, which collided beams of protons against fixed targets. After many drawbacks and difficulties, on March 1, 1972 the machine accelerated a beam of protons to the design energy of 200 GeV, but more power supplies could be added later to increase the accelerator's peak energy to more than 400 GeV. Then, in 1977 a new meson was found whose constituents were identified to be a new quark, which was called *bottom*, and its antiquark. Thus, this resulted in the discovery of the lightest quark of the third family of fermions, along with its antiquark, only one year after the discovery of the tau lepton. However, the discoveries of the top quark and its antiquark and of the tau neutrino, also from the third family, still had to wait another 18 and 23 years, respectively, as we will see below.

Fermilab: The Tevatron

In 1983, another much more powerful synchrotron was built in the same tunnel as the Main Accelerator. This was a proton-antiproton ($p\,\bar{p}$) collider, reaching energies of almost 1 TeV (1000 GeV) per beam, 1.96 TeV in total, which is why it was called Tevatron (Fig. 6.10). In the construction of this accelerator, technology based on superconducting magnets was used for the first time, which requires ultra low temperatures close to absolute zero (0 K). The Tevatron was running from 1983 to 2011, and became the most powerful accelerator in the world from 1985 until the LHC went into operation at CERN, in November 2009. The main achievement of the Tevatron was the 1995 discovery of the top quark and its antiquark. In addition, in the year 2000 the tau neutrino was also discovered using the Tevatron, and with it the entire third family of fermions was finally detected.

Fig. 6.8 The Robert R. Wilson Hall, main building of the Fermi National Accelerator Laboratory (Fermilab). It was named after its first director, who designed the building as well as many sculptures around the site. *Credit* Courtesy of Fermilab

Fig. 6.9 Enrico Fermi (1901–1954) around 1943-49. He was one of the most prominent physicists in the twentieth century. He excelled in both theoretical and experimental Physics. Particles with half-integer spin are named "fermions" because their behavior is described using his Fermi–Dirac statistics. He was the creator of the world's first nuclear reactor, the Chicago Pile-1, and was called the "architect of the atomic bomb". Fermi was awarded the 1938 Nobel Prize in Physics for his work on induced radioactivity by neutron irradiation and for the discovery of nuclear reactions brought about by slow neutrons. He has also a famous paradox as regards extraterrestrial civilizations. *Credit* Courtesy of Wikimedia Commons

Fig. 6.10 Aerial view of Fermilab's accelerator complex, where the Wilson Hall can be distinguished. The Main Injector Ring, 3.3 km long, is in the foreground and the Tevatron, 6.3 km long, is in the background. This proton-antiproton collider was the most powerful accelerator in the world between 1985 and 2009. It was shut down in 2011. *Credit* Courtesy of Fermilab

Fermilab: The Main Injector Ring

Currently, Fermilab continues performing high-energy experiments using the Main Injector Ring, 3.3 km long (Fig. 6.10), which was completed in 1999 and accelerates protons until an energy of 120 GeV. These protons are used in a number of fixed-target experiments, like the E-906/SeaQuest which tries to measure the contributions of antiquarks (from the quark-antiquark sea) to the structure of protons and neutrons. Furthermore, Fermilab is involved in several neutrino experiments, including the future Deep Underground Neutrino Experiment (DUNE), under construction, and it also designs and builds accelerators, among other scientific activities. The asteroid "11,998 Fermilab" is named in honor of the laboratory.

KEK: The Synchrotrons PS, TRISTAN, KEKB, SuperKEKB and the KEK Linac

In 1971 the National Laboratory for High Energy Physics (KEK) was created in Tsukuba, Japan, which is the largest particle physics laboratory in that country. In 1976 the Proton Synchrotron (PS), produced a proton beam with 8 GeV, as designed, although later the protons reached 12 GeV. It was shut down in 2007. In 1986 an electron–positron (e^-e^+) collider was built, the Transposable Ring Intersecting STorage Accelerator in Nippon (TRISTAN) that accelerated both electron and positron beams up to 30 GeV. This accelerator operated until 1995, when it was converted into the KEKB e^-e^+ collider, a so-called B-factory designed to produce B mesons copiously for the Belle experiment, which started in 1998. The latter accelerator was super-seded by its upgrade, the SuperKEKB collider, which had its first particle collisions in 2018. It produces B mesons for the Belle II experiment, which is an upgrade of the Belle experiment. The electrons and positrons are previously accelerated in the linear accelerator KEK Linac (Fig. 6.11), 600 m in length, which is used to inject 8 GeV electrons and 3.5 GeV positrons to KEKB. The latter has been upgraded recently to serve as pre-accelerator for the SuperKEKB collider.

KEK: Neutrino experiments and J-PARC
Among other tasks, the PS provided the proton beam to a neutrino beam line for the "KEK to Kamioka" (K2K) experiment. To be precise, a beam line to drive neutrinos into the Super-Kamiokande neutrino detector, which is about 250 km away from KEK, was established and the neutrino oscillation experiment K2K was conducted from 1999 to 2004. In 2005 the KEK laboratory was enlarged with the Tokay campus and, at present, it is a very large facility and accelerator complex where a whole variety of research lines are being carried out. In 2009 the construction of the Japan Proton Accelerator Research Complex (J-PARC) in the Tokay campus was concluded. J-PARC uses high intensity proton beams of 50 GeV, produced in a synchrotron called Main Ring (MR), to create high intensity secondary beams of hadrons and neutrinos. Using the latter beams, since 2009 the neutrino oscillation experiment "Tokai to Kamioka" (T2K) has been conducted for analysis at the

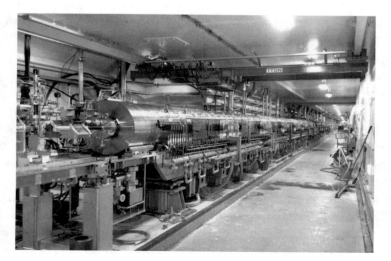

Fig. 6.11 The multi-purpose linear accelerator KEK Linac, 600 m long, used for supplying electron and positron beams with a maximum energy of 10 GeV. *Credit* Courtesy of KEK

Kamioka laboratory, 295 km to the west. We will return to this experiment in Chap. 7.

Before finishing this chapter, let us take a closer look at CERN, which is the largest laboratory of Particle Physics in the world. In so doing we will expand some of the information that was given above.

6.4   CERN

In 1949, while Europe was still recovering from World War II, a few physicists realized the need of creating an European laboratory for the study and development of Nuclear Physics. For one thing, European science had to be rebuilt and fundamental research had to be revived, but, in addition, the constant brain drain to the USA had to be halted. The brain drain of prominent physicists, that had already begun before the war (Albert Einstein, Enrico Fermi, and Bruno Rossi, among others), had dramatically increased during and after it. Hence, it was urgent to provide a means to unite European physicists as well as to equip Europe with modern research facilities whose costs would be shared among the participating research institutions in the different countries.

With this in mind, French physicist Louis de Broglie became the first prominent scientist to propose the creation of a multi-national European laboratory for Nuclear Physics research. He had been the recipient of the 1929 Nobel Prize in Physics for his discovery of the wave nature of electrons, and then, twenty years later, he was putting forward this official proposal at the European Cultural Conference, in Lausanne in December 1949. Two years later, at a meeting of UNESCO in Paris, in December 1951, the first resolution concerning the establishment of the Conseil Européen pour la Recherche Nucléaire, with the acronym CERN, was adopted. And only two months later, the representatives of 12 countries signed an agreement establishing the provisional Council (Conseil, in French), agreement that gave rise to the formal CERN Convention 16 months afterwards. The CERN Council changed its name some time later to European Organization for Nuclear Research, but it kept the acronym.

The sixth session of the CERN Council took place in Paris, 29 June – 1 July 1953. It was there that the convention establishing the Organization was signed, subject to ratification, by 12 States. The ratification still took several months and finally CERN was officially created in 1954. Its mission: to uncover what the Universe is made of and how it works. Moreover, CERN's

Convention explicitly included a *Science for Peace* statement: *The Organization shall have no concern with work for military requirements and the results of its experimental and theoretical work shall be published or otherwise made generally available.* In July 1955, CERN's Director-General, Felix Bloch, laid the first foundation stone.

At present, CERN brings together physicists from all over the world. It comprises 23 Member States, and there are also eight Associate Member States, including India and Pakistan. Furthermore, Japan, the Russian Federation, the USA, as well as the European Union, JINR and UNESCO have "Observer" status. CERN's main area of research is Particle Physics although it is also quite active in Nuclear Physics. In addition, since 1995 a major effort is being made in the study of antimatter atoms, as we will see in detail in Chap. 8. Over 600 research centers and universities from more than 70 countries around the world use CERN's facilities. This amounts to more than 12,000 scientists of 110 nationalities, and funding agencies from both Member and Non-Member States are responsible for financing, constructing, and operating the experiments on which they collaborate.

Near the CERN entrance, the Globe of Science and Innovation, a symbol of planet Earth, was placed in 2004 (Fig. 6.12). It harbors the permanent

Fig. 6.12 The Globe of Science and Innovation, a wooden structure 27 m high and 40 m in diameter, is a symbol of planet Earth located near CERN entrance since 2004. It harbors the permanent exhibition "Universe of Particles". In its proximity lies the 15-tonne steel sculpture "Wandering the immeasurable" by the artist Gayle Hermick as of 2014. The sculpture, 11 m high, is shaped like a giant 37 m ribbon, and pays homage to 396 great discoveries in Physics, Astrophysics, and Mathematics through the ages, which are laser engraved on its surface in their language of origin. *Credit* Courtesy of CERN

exhibition *Universe of Particles*. Ten years later the steel sculpture *Wandering the immeasurable*, by the artist Gayle Hermick, was inaugurated in its proximity (Fig. 6.12). The sculpture, 11 m high, is shaped like a giant 37 m ribbon, and pays homage to 396 great discoveries in Physics, Astrophysics, and Mathematics through the ages, which are laser engraved on its surface in their language of origin. Moreover, since 2004 another piece of art adorns the CERN campus in the square between the Hostel buildings 39 and 40. It is a bronze statue of Lord Shiva, the destroyer of the Universe according to Hinduism, which was a gift from the Department of Atomic Energy in India to celebrate its 40-year association with CERN (Fig. 6.13).

Among its many achievements, CERN also has the merit of being the cradle where the World Wide Web (www) was conceived and developed. Indeed, it started as a CERN project, led by Tim Berners-Lee in 1989, whose purpose was to facilitate the transfer and exchange of data among scientists from all over the world through the then emerging Internet. Berners-Lee defined the basic concepts of URL, http and html, and wrote the first software for a browser and a server. The first web page was activated in 1991 (it is still possible to visit it), and in 1993 CERN decided to make the www available, free of charge, to Internet users worldwide.

In Particle Physics research, CERN has built the three largest accelerators to date and also the most powerful ones, with the exception of Fermilab's Tevatron, only surpassed in power by the LHC.

The first of these enormous machines was the Super Proton Synchrotron (SPS), with a 6.9 km ring inside an underground tunnel across the border between Switzerland and France. As explained in the previous section, it came into operation in 1976, its detectors discovered the $W^{\pm}$ and Z^0 bosons in 1983, and still is in use today, reaching a maximum energy of 450 GeV per beam. It has been given many uses, both in experiments with a fixed target and as a collider, since it is very versatile. Indeed, in the SPS ring the four stable particles have been accelerated: electrons, positrons, protons, and antiprotons; as well as nuclei of oxygen, sulfur, and lead. In addition, since the advent of the LHC, the SPS is used to supply the required protons and lead nuclei (see Fig. 6.14). An interesting fact is that the Tevatron, being slightly smaller in size than the SPS (6.3 vs. 6.9 km), was nevertheless twice as powerful (900 vs. 450 GeV of energy per beam). This was possible because the Tevatron was equipped with more than 1000 superconducting electromagnets, whereas the SPS counts with a slightly larger number of conventional electromagnets (1317), which generate much less intense magnetic fields than the superconducting ones.

Fig. 6.13 Statue of Lord Shiva, the Hindu deity who will eventually annihilate our Universe. His depiction performing the Cosmic Dance is named Nataraja. It is a well known sculptural motif in India, one of the finest illustrations of Hindu art, often used as a symbol of Indian culture. The statue was donated in 2004 to CERN by the Department of Atomic Energy in India as a gift to celebrate the 40-year association between the two institutions. *Credit* Courtesy of Kenneth Lu and Wikimedia Commons (Creat. Comm. 2.0 license)

The next accelerator built at CERN was the gigantic Large Electron Positron Collider (LEP), 26.7 km in length. As already mentioned, it was also installed in an underground tunnel, but much deeper than the SPS, at about 100 m in depth, and also across the Franco-Swiss border (around three quarters of the tunnel are located in French territory). It started operation in 1989 and for seven years ran at about 100 GeV (half per beam), producing some 17 million Z^0 bosons, which have a mass of 91 GeV. Afterwards, its technology was improved to reach 209 GeV and thus be able to produce pairs of W^+ and W^- bosons, whose mass is 80 GeV. The LEP collider

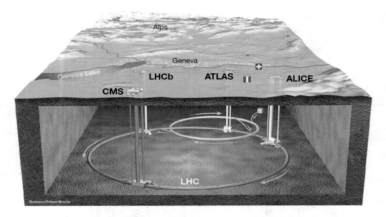

Fig. 6.14 Sketch of the Large Hadron Collider (LHC), hosted in a 26.7 km long tunnel about 100 m underground across the border between Switzerland and France, near Geneva. One can see the location of the four main detectors: ATLAS, CMS, ALICE and LHCb, and also depicted is the ring of the Super Proton Synchrotron (SPS), which injects protons and lead nuclei into the LHC. *Credit* Courtesy of CERN

succeeded in making highly accurate measurements on already known results, which proved that there are only three types of light neutrinos, and consequently only three families of fermions. It failed to discover the Higgs boson, although it came quite close, and was closed down in November 2000 in order to dismantle it and proceed with the construction of the LHC in the same tunnel, as pointed out before.

Now the question arises as to why the LEP, being much larger than the SPS, was nevertheless much less powerful, reaching no more than a quarter of the energy of the latter. The answer is that electrons (and positrons) radiate much more electromagnetic energy than protons (and antiprotons) when following curved trajectories. As a result, circular accelerators are much less efficient for electrons than for protons due to the so-called synchrotron radiation whereas, on the contrary, linear accelerators are much more efficient for electrons than for protons, as the latter are heavier and require more energy to be accelerated.

Another related question is why the CERN scientists decided to embark on such a feat of engineering and cutting-edge technology only to end up building a much less powerful accelerator than the previous one. The reason is that e^+e^- collisions are very clean, because these particles are elementary and do not present any structure, so these collisions allow very precise measurements on the resulting products. The collisions between hadrons, on the contrary, occur between two multitudes of particles. Consequently, the energy of the collision is not known with accuracy, as the energy of the hadron is distributed among its constituents. Moreover, the large amount of particles resulting from hadronization makes it very difficult to analyze the final products and to obtain precise measurements.

The last of the huge accelerators built by CERN is the Large Hadron Collider (LHC). Installed inside the 26.7 km tunnel that housed the LEP (Fig. 6.6), and using the superconducting electromagnet technology inherited from the Tevatron, the LHC stands as the largest and most powerful accelerator that exists today (Fig. 6.14). It is equipped with 9593 superconducting electromagnets of diverse kinds. These include 1232 main dipole magnets 15 m long which bend the beams, and 392 main quadrupole magnets, between five and seven meters long, which focus the beams. And just before the collision another type of magnet is used to "squeeze" the particles closer together to increase the chances of collisions. The LHC operates mainly as a proton-proton (p p) collider, but it also accelerates lead (Pb) nuclei, which it makes collide with other Pb nuclei, or with protons. In either case, as shown in Fig. 6.5, the beams start their journey in small linear accelerators, Linac 4 for protons (which has replaced Linac 2 in 2020) and Linac 3 for ions, entering the PS Booster, which brings them to a higher energy. From there they continue their way in the acceleration chain being subsequently injected and accelerated, first in the Proton Synchrotron (PS), then in the Super Proton Synchrotron (SPS), and finally in the LHC ring.

The LHC started functioning in September 2008, but it suffered a series of explosions right at the beginning, due to faulty connections between the magnets, causing it to shut down for over a year. Due to this, the LHC actually began functioning in November 2009. A curious anecdote about the accident is that when the technicians inspected the tunnel, after the explosions, they found puddles on the ground all along it, resulting to be the air that had liquefied due to the very low temperatures. This is so because the superconducting electromagnets only function properly at a temperature of around, or below, 1.9 K (-271.25 °C), which is attained with the aid of liquid helium. In fact, this temperature is inferior to the lowest temperature one finds in outer space, bathed by the CMB radiation at 2.7 K (-270.4 °C).

The purpose of the LHC is twofold: to test the Standard Model of Particle Physics at increasingly levels of accuracy, and to search for new physics beyond the Standard Model in order to validate the predictions of some of the current theoretical models. This includes creating so far unknown particles, such as dark matter particles, SUSY (supersymmetric) particles, and Kaluza-Klein particles, the latter in order to probe extra dimensions. Along the LHC ring there are seven particle detectors, of which the four principal ones are ATLAS, CMS, ALICE and LHCb (Figs. 6.5 and 6.14). The ATLAS detector

(Fig. 6.15) is the largest, with 25 m in diameter, 46 m in length and 7,000 tons of weight; followed by the CMS detector (Fig. 6.15), 15 m in diameter, 22 m in length and weighing 12,000 tons. Both the ATLAS and CMS experiments were involved in the discovery of the Higgs boson in 2012 (Fig. 6.16). The detectors TOTEM, MoEDAL and LHCf, are much smaller than the others and serve for very specialized research.

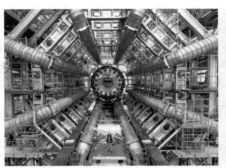 

Fig. 6.15 The detectors ATLAS (left) and CMS (right) of the Large Hadron Collider (LHC) at CERN. They were involved in the discovery of the Higgs boson. *Credit* Courtesy of CERN

Fig. 6.16 Event recorded by the CMS detector of a proton-proton collision with 8 GeV of energy at the LHC, CERN, on May 27, 2012. It shows the by-products from the decay of the Higgs boson into a pair of Z^0 bosons, one of which decays giving an electron–positron pair (green lines) and the other Z^0 decays giving a muon-antimuon pair (red lines). *Credit* Courtesy of CERN

The Principal LHC Detectors

Along the ring of the Large Hadron Collider (LHC), there are four main detectors: ATLAS, CMS, ALICE and LHCb. Let us say a few words about them and also about MoEDAL.

ATLAS

A Toroidal LHC ApparatuS is the largest, general purpose detector at the LHC. It was designed to discover heavy particles which are not observable using less powerful accelerators, and to search for evidence of phenomena beyond the Standard Model. This includes SUSY particles, Kaluza-Klein particles to probe extra dimensions, and dark matter particles. ATLAS was one of the two LHC experiments that discovered the Higgs boson.

CMS

Compact Muon Solenoid is the other general purpose detector at the LHC. Like the ATLAS experiment, the CMS experiment was also involved in the discovery of the Higgs boson and it investigates a wide range of phenomena beyond the Standard Model, including SUSY, extra dimensions, and dark matter particles.

ALICE

A Large Ion Collider Experiment was designed to study heavy-ion collisions using lead nuclei with total energy up to 5.02 TeV. This translates into temperatures and energy densities corresponding to a quark–gluon plasma, in which quarks, antiquarks and gluons move freely. This fifth state of matter existed in the first fractions of a second after the Big Bang, before quarks, antiquarks and gluons bound together to form hadrons. The study of the quark–gluon plasma can shed light on the origin of the hadron masses obtained through the strong interactions, which are much larger than the masses obtained through the Higgs mechanism, as explained in Chap. 2, Sect. 2.5.4.

LHCb

Large Hadron Collider beauty is an experiment specialized in the physics of the bottom b quark, also dubbed "beauty". This quark and its antiquark are involved in processes that do not preserve the CP symmetry between particles and antiparticles (we will see all the details in next chapter). In order to study this phenomenon, called CP violation, the LHCb experiment was designed to obtain more data about it in the B meson system (mesons containing a bottom quark or antiquark). This experiment is directly related to the problem of the asymmetry between matter and antimatter in the Universe, which is the subject of Chap. 7.

MoEDAL

Monopole and Exotics Detector at the LHC is an experiment whose prime goal is to search for magnetic monopoles – hypothetical particles with magnetic charge - or dyons, which are equally hypothetical particles but with both electric and magnetic charges. MoEDAL is located in the same LHC cavern as the LHCb experiment.

In the first stage of operation (Run 1), that started November 20, 2009, the LHC attained 600 million $p\,p$ collisions per second, reaching a maximum energy of 8 TeV (8000 GeV), half per beam. Already on November 30, ten days after the first proton beams began circulating for the first time after the accident, the LHC achieved 1.18 TeV per beam. Therefore, the LHC became the highest energy particle accelerator in the world, beating the Tevatron's previous record of 0.98 TeV per beam, held for eight years. Only four months later, on March 30, 2010, the LHC set a new record by colliding proton beams with a total energy of 7 TeV. The first round of $p\,p$ collisions ended on November 4, 2010, and was followed by another round of collisions but with lead nuclei, i.e. Pb Pb collisions, for four weeks. This allowed the ALICE experiment to study the quark-gluon plasma for the first time. The historic discovery of the Higgs boson (Figs. 6.16 and 6.17), which was made by the ATLAS and CMS collaborations, was announced in the CERN Auditorium on July 4, 2012, by the ATLAS spokesperson, Fabiola Gianotti (Fig. 6.18), in a major media event. She is currently the Director General of CERN as of 1st January 2016, the first woman to be appointed this position in 60 years since the establishment of the laboratory.

On February 13, 2013, the LHC was shut down for a two-year upgrade. In the second stage of operation (Run 2), from April 5, 2015 to December 3, 2018, more than 1000 million $p\,p$ collisions per second were achieved and the LHC was able to operate with a total maximum energy of 13 TeV, although it also worked with lower energies of up to 8.16 TeV when it collided Pb nuclei. Furthermore, in 2016 the round of $p\,p$ collisions was followed for the first time by proton-lead p Pb collisions for a period of four weeks. Run 2 saw no new discoveries, but allowed researchers to gain increased confidence in the

Fig. 6.17 Peter Higgs visiting the cavern of the CMS experiment at the LHC, CERN. He stands in front of the detector, which was open due to maintenance work. The CMS and ATLAS experiments discovered the Higgs boson in 2012. *Credit* Courtesy of CERN

Fig. 6.18   The ATLAS collaboration spokesperson, Fabiola Gianotti, at the CERN Auditorium, presenting on July 4, 2012, the discovery of a particle consistent with the long-sought Higgs boson in the mass region around 125–126 GeV. Fabiola Gianotti is currently the Director General of CERN since January 1, 2016. *Credit* Courtesy of CERN

assumption that the particle discovered in 2012 was actually the Standard Model Higgs boson.

On December 10, 2018, the LHC was shut down again along with the whole CERN accelerator complex for another "two-year" upgrade, which in this case will take slightly more than two years for the PS and other accelerators, but more than three years for the LHC, as Run 3 has been rescheduled to start around February 2022. Once that third stage is completed, an even longer shut down will take place at CERN for about three years, during which many major upgrades and improvements will be made to the LHC in order to increase its luminosity[2] by a factor of 10, which will substantially increase the number of collisions and data collected by the researchers. The accelerator will even be renamed as High Luminosity Large Hadron Collider (HL-LHC), and should be operational around 2028. It will produce at least 15 million Higgs bosons per year, whereas the LHC produced about three millions in 2017. This project is led by CERN with the support of an international collaboration of 29 institutions in 13 countries, including the USA, Japan, and Canada.

---

[2]The luminosity of a beam is the number of particles per unit of surface swept and unit of time. Hence, the number of collisions produced when making a beam hit a fixed target, or another beam, is directly proportional to its luminosity.

6.5   Beyond the LHC

There are several projects underway to build other accelerators of larger size than the LHC, both linear and circular, although none of them has been confirmed. At CERN alone there are three such projects with the accelerator acronyms CLIC, FCC and HE-LHC. The Compact Linear Collider (CLIC) would be a linear electron–positron collider built at CERN with a length between 30 and 50 km and collision energies up to 3 TeV. It would start running by the time the HL-LHC had finished operations, around 2035. The Future Circular Collider (FCC), in turn, would be a 100 km synchrotron with a collision energy of 100 TeV, being built either around CERN or in China. Alternatively, a High Energy Large Hadron Collider (HE-LHC) could replace the HL-LHC in the same tunnel, reaching a collision energy of 27 TeV and delivering a luminosity three times higher than the HL-LHC.

Another important proposal is the International Linear Collider (ILC), a linear electron–positron collider consisting of superconducting cavities (the devices supplying the electrical power to the beams) with a total length of approximately 31 km. The electron and positron energy will be up to 500 GeV with the option to upgrade it to 1 TeV. In 2013, the Kitakami Mountains in Japan were chosen the best site to build the ILC. In this project, nearly 300 laboratories and universities around the world are already involved; more than 700 people working on the accelerator design, and another 900 people on the detector development.

7

Matter–Antimatter Asymmetry

7.1 Matter–Antimatter Asymmetry in Astrophysics

The asymmetry between the amounts of matter and the amounts of anti-matter in our observable Universe is overwhelming and constitutes one of the most striking and fascinating enigmas of Particle Physics and Cosmology. However, the exact magnitude of this asymmetry is not clearly and defini-tively deduced from astronomical observations, and their estimates depend on the theoretical models used to describe the physics of the Early Universe, in which this asymmetry would have been generated. For these reasons there is discrepancy among the scientists; and although the prevailing opinion is that the primordial antimatter disappeared in its entirety during the first instants of existence of the Universe, there is no scarcity of those who advocate that it was not necessarily so and that small islands of primordial antimatter could have survived in our observable Universe. In this event, they would exist in the form of structures such as gas clouds, stars, black holes and even galaxies, all consisting solely of antimatter.

When we look at the skies on a starry night, far from the urban lights, we witness an unparalleled spectacle. Myriads of luminous little dots surround us in all directions, concentrating particularly on a strip that our ancestors called the Milky Way, since it resembles a path of spilled milk. Today we know that this strip is nothing but the disk of the galaxy that hosts us, as viewed from the inside, which is where our Solar System is located. We understand that those luminous points that we observe correspond to astronomical objects, mainly stars, in addition to the six planets of the Solar System that can be

© Springer Nature Switzerland AG 2021
B. Gato-Rivera, *Antimatter*,
https://doi.org/10.1007/978-3-030-67791-6_7

seen with the naked eye. If in those moments we focused our attention on any little dot, and we asked ourselves if the shining body is made up of matter or antimatter, what would be the answer? Could we somehow deduce if the atoms and nuclear reactions which emitted that light were atoms of matter and reactions between their nuclei or, on the contrary, they were atoms and nuclear reactions of antimatter?

From what we know so far, and unless we get unexpected surprises from the experiments being performed at CERN's Antimatter Factory, the light emitted by an antimatter atom should be identical to the light emitted by its matter counterpart[1]; and the same should be said about the photons that would be produced in the nuclear reactions deep within the antistars, if they were to exist. Consequently, it is not possible to distinguish whether a star is made of matter or antimatter judging by the light, or any other electromagnetic radiation, that reaches us from it. Paul Dirac himself, in his speech at the 1933 Nobel Prize ceremony, entitled *Theory of electrons and positrons*, already underlined this curious aspect of antimatter in his final words:

> If we accept the view of complete symmetry between positive and negative electric charge so far as concerns the fundamental laws of Nature, we must regard it rather as an accident that the Earth (and presumably the whole solar system), contains a preponderance of negative electrons and positive protons. It is quite possible that for some of the stars it is the other way about, these stars being built up mainly of positrons and negative protons. In fact, there may be half the stars of each kind. The two kinds of stars would both show exactly the same spectra, and there would be no way of distinguishing them by present astronomical methods.

7.1.1 Neutrinos or Antineutrinos: The Key Factor

Ironically, the stars carry a label that was not known at the time of Dirac's speech, an unmistakable sign that shows whether they are made of matter or antimatter, and that would allow us to differentiate between the two possibilities. The problem is that with our present instruments we do not have the possibility to read those labels, except for the Sun. Let us take a closer look at this issue.

In the deepest layers in the interior of the stars, the nuclei of the atoms fuse together producing larger nuclei. This fusion requires enormous temperatures, in our Sun around 15 million degrees (15,600,000 K), so that the

---

[1] Identical for all practical purposes, although there should be some extremely small discrepancies caused by the violation of the CP symmetry that we will discuss below in Sect. 7.2.4.

protons can overcome the electrostatic repulsion that separates them and can get close enough to be held together by the strong force, much more powerful than the electric one, but with a very short range. Even so, an additional effect from the realm of quantum mechanics has to come into play for the fusion actually to take place. This is *quantum tunneling*, that is only possible, although with very low probability, if the protons approach each other at very short distances, which requires also enormous temperatures. Without quantum tunneling,[2] the temperature necessary to overcome the electrostatic repulsion between protons would be much higher than that of the Sun, so that nuclear fusion would only have taken place in much denser stars than the main sequence stars to which the Sun belongs.

The primary nuclear reaction within stars is often described as the combustion of primordial hydrogen, created in the early Universe, resulting in the production of helium. But, in reality, at the high temperatures present in the stellar cores, protons and electrons cannot bind together to form atoms; they form a plasma instead, leaving atoms relegated to the outer layers with more tolerable temperatures. As explained in Chap. 2, Sect. 2.5.3, the set of nuclear reactions in "hydrogen burning" stars, like the Sun, in which helium nuclei are produced out of protons, is called the *proton–proton chain reaction* (Fig. 7.1). These reactions start with the fusion of two protons ($p + p$) resulting in a deuterium nucleus ($p + n$), called deuteron, plus one positron e^+ and one electron neutrino ν_e, as shown in (2.12), that we repeat here for convenience:

$$p + p \rightarrow p + n + e^+ + \nu_e. \tag{7.1}$$

This process is caused by the weak interactions mediated by the W^+ boson and has a very low probability. Therefore, the union of two protons producing one deuteron, which is the first step for nuclear fusion, is due to the combined action of two processes, each with very low probability: the quantum tunneling allowing the protons to approach each other at the very short range of weak forces (10^{-16} m), and the weak interactions themselves. As a consequence, each individual proton in the hydrogen burning stars is bouncing around for about five thousand million years before it has the chance to overcome the electrical repulsion of another proton and get involved in the proton-proton chain reaction culminating in the production of helium nuclei.

---

[2]Quantum tunneling is a phenomenon which becomes relevant at the nano scale (10^{-9} m) and smaller ones. It enables subatomic particles and atoms to cross barriers without having the required energy to overcome them. It is also of vital importance for life processes.

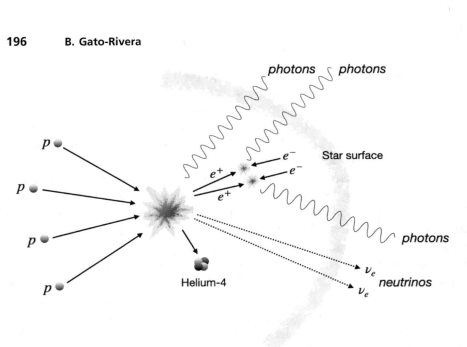

Fig. 7.1 Illustration of the "proton-proton chain reaction" in the stellar cores, in which four protons—four hydrogen nuclei—are fused to produce one nucleus of Helium-4. This process takes several steps and is accompanied by the emission of two positrons e^+, two electron neutrinos ν_e and two or three photons. The positrons annihilate themselves very quickly with the electrons e^- of the surrounding plasma, producing more photons. Therefore, from the particles originating in the nuclear reactions within stars, only photons and electron neutrinos are emitted to outer space, although the photons can take hundreds of thousands of years to reach the outer layers of the stars before being released

The second step, much faster than the previous, is the fusion of a proton with one deuteron producing a nucleus of ^{3}He and one photon. Then, two of these reactions allow to fuse two ^{3}He nuclei resulting finally in one nucleus of ^{4}He and two protons. Moreover, once there are ^{4}He nuclei available—α particles consisting of two protons and two neutrons—they can also participate in the creation of more of these nuclei through two different sets of reactions, although the fusion of two ^{3}He nuclei is the process that takes place about 85% of the time. Furthermore, an alternative way to transform protons in helium nuclei in the stellar cores is a cycle of nuclear reactions called CNO (carbon–nitrogen-oxygen), which is predominant in stars more massive than the Sun. In this cycle the nuclei of these three elements act only as catalysts but the net result is the same as in the proton-proton chain reaction, including the emission of positrons, photons, and electron neutrinos.

In summary, what happens in the nuclear furnaces of hydrogen burning stars is that four protons, i.e. four nuclei of hydrogen ^{1}H, are fused resulting in one nucleus of helium ^{4}He accompanied by the emission of two positrons e^+, two electron neutrinos ν_e and, at least, two photons γ, as shown in Fig. 7.1. Hence, the Sun and all hydrogen burning stars produce enormous amounts of antimatter in the form of positrons as well as enormous amounts of electron neutrinos and photons. The latter get stuck inside the plasma, and may need hundreds of thousands of years to emerge in the outer layers of the star and drift away into outer space. By contrast, the positrons annihilate themselves very quickly with the electrons of the plasma, producing two or three more photons for each annihilation. In fact, some estimates indicate that around 10% of the visible daylight we receive today on Earth has its origin in the photons that were produced in the electron–positron annihilation within the Sun hundreds of thousands of years ago.

As for the neutrinos, they are immediately ejected from the star at speeds close to the speed limit c, passing through the star and everything ahead of them, including the planets they encounter on their way. Indeed, about eight minutes after they are created within the Sun, an astonishingly large number of solar neutrinos reaches the Earth,[3] resulting in a flux of about 70,000 million solar neutrinos per cm^2 every second.

Now, in a star of antimatter, that would exist if a sufficient amount of primordial antiprotons had survived the Great Annihilation at the beginning of the Universe, the analogous *antiproton-antiproton chain reaction* would start with the fusion of two antiprotons $(\overline{p}+\overline{p})$, resulting in one antideuteron $(\overline{p}+\overline{n})$ plus one electron e^- and one electron antineutrino $\overline{\nu}_e$:

$$\overline{p} + \overline{p} \rightarrow \overline{p} + \overline{n} + e^- + \overline{\nu}_e. \tag{7.2}$$

And the net result of the nuclear reactions burning antihydrogen would consist, therefore, of the fusion of four antiprotons, resulting in a nucleus of antihelium $^4\overline{\text{He}}$ (an antialfa $\overline{\alpha}$ particle) accompanied by the emission of two electrons e^-, two electron antineutrinos $\overline{\nu}_e$ and, at least, two photons γ, as shown in Fig. 7.2. The electrons would be annihilated immediately with the positrons of the antiplasma, giving two or three more photons for each annihilation, and the antineutrinos would be expelled from the antistar at speeds close to the speed limit c.

The photons would undergo the same fate as in the stars of matter, so there would be no way to differentiate a star from an antistar judging by the light emitted by the "nuclear versus antinuclear" reactions, and the same can

---

[3]The sunlight takes eight minutes, approximately, to travel from the outside layers of the Sun to the Earth, so this should be about the same time needed for the neutrinos coming from the deep layers, as they travel at nearly the same speed c as the light and they traverse the Sun very quickly.

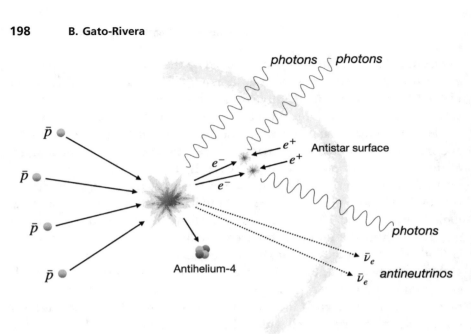

Fig. 7.2 Illustration of the "antiproton-antiproton chain reaction", that would take place inside antistars, if they existed. Four antiprotons would become fused to produce one nucleus of Antihelium-4 accompanied by the emission of two electrons e^-, two electron antineutrinos $\bar{\nu}_e$ and photons. The electrons would annihilate with the positrons e^+ of the antiplasma, producing more photons. Hence, from the particles originating in the nuclear reactions within antistars, only photons and electron antineutrinos would be emitted to outer space, although the photons would take hundreds of thousands of years to reach the outer layers of the antistar before being released

be said of the light emitted by the atoms or antiatoms in the outer layers of the stars or antistars, as pointed out above. Therefore, the labels that would unambiguously indicate whether a star is made up of matter or of antimatter are written with electron neutrinos or antineutrinos, respectively. But these labels are beyond the reach of our current rudimentary neutrino telescopes since these do not have the capacity nor the resolution to observe them, except in the case of the Sun, for which the "neutrino label" is observed. It should be noted, however, that the Sun and most stars contain also radioactive nuclei that release antineutrinos, although their quantities are completely insignificant in comparison with the quantities of neutrinos released in the nuclear reactions. In the case of the Sun, the difference amounts to eight orders of magnitude, at least in theoretical estimates, because up to date the solar antineutrinos have never been observed.

7.1.2 Astronomical Observations: DGRB and CMB

Whether or not primordial antimatter exists, and regardless of whether it has given rise to antimatter structures, it has been known for several decades that the dominance of matter over antimatter is overwhelming, at least in the observable part of our Universe. This means, in particular, that it could well be that all the luminous little dots that we admire in a starry night are constituted of ordinary matter. As already explained in Chap. 4, Sect. 4.6.1, the relevant astronomical observations, which in principle could provide information about the existence of primordial antimatter in the Universe, pertain to three areas of research: the *Diffuse Gamma Ray Background* (DGRB), the *Cosmic Microwave Background* (CMB), and the *Cosmic Rays*.

We have already covered the subject of the Cosmic Rays in Chap. 5. We saw that, as far as antimatter is concerned, only antiprotons and positrons have been found in the primary cosmic rays in a conclusive way, and in very small amounts compared to the abundance of matter particles. Now, it turns out that the observed fluxes are not in agreement with the estimates of their secondary production, either in collisions of cosmic rays in the interstellar medium or in known astrophysical processes. Indeed, there is an excess of antiparticles in such fluxes, especially in the case of positrons, so one might wonder whether this is a signal that primordial antiparticles actually exist. However, careful analysis seems to indicate that this excess might be a signal of dark matter decay or annihilation, or alternatively, that some astrophysical objects or processes are much more powerful antimatter sources than previously believed. In this respect, neutron star pulsars and massive black holes are the prime suspects.

Hence, there are no clues about the possible existence of primordial antiparticles created in the first instants of the Universe because the quantities of observed antiprotons and positrons may be explained in terms of secondary production. In other words, it could well be that all the antimatter particles in the primary cosmic rays hitting our atmosphere were secondary antimatter particles. But it could also be that a small percentage of those cosmic-ray antiparticles were actually primordial. In either case, we will have to wait still some years to know the answer. In fact, the waiting time could be drastically shortened if the detections of Antihelium-3 and Antihelium-4 nuclei by the AMS collaboration were confirmed, which could occur around 2024. We will come back to this issue in the next section, where we will discuss the possibility of antimatter structures in the Universe, and even in our galaxy.

The Diffuse Gamma Ray Background, DGRB, is interpreted as the accumulation of γ ray emissions from sources that are not bright enough to be detected individually. This means that the sources may not be bright, or they may be so far away that we cannot resolve them with our telescopes and γ ray detectors. Its exact composition is unknown, although it is believed to have a galactic component, that is, emissions within our own galaxy, and also an extragalactic component.

The Cosmic Microwave Background, CMB, in turn, can be considered as a fossil radiation, since it was released in the Early Universe. As was explained in Chap. 3, it consists of photons in the microwave range, centered on the wavelength $\lambda = 1.9$ mm, and its power spectrum corresponds to the black body radiation with a temperature of 2.7 K (see Fig. 3.6). These photons were created mainly in the Great Annihilation between matter and antimatter and remained confined for 380,000 years, trapped in the plasma of electrons, protons and α particles, bouncing among them. Then, when the temperature of the Universe had dropped enough, the electrons combined with the protons and α particles to form hydrogen and helium atoms, causing the Universe to become transparent to photons, which were released as a consequence. Moreover, it cannot be discarded that in that plasma there was also some primordial antimatter in the form of positrons, antiprotons and $\overline{\alpha}$ antiparticles, although this is a matter of debate and even controversy. If this were the case, then at the same temperature the positrons would have combined with the antiprotons and $\overline{\alpha}$ antiparticles to form antihydrogen and antihelium atoms. In addition, if the matter and antimatter atoms were separated in different regions, some annihilations would have occurred in the borders between them.

As for the possibility of observing signals in the DGRB gamma radiation that indicate the presence of antimatter in the Universe, in principle this would be possible through the electromagnetic radiation coming from the borders or boundaries between regions of matter and regions of antimatter. These borders, which are not observed, would exhibit an enormous production of energy, mainly γ rays, due to the annihilation between particles of matter and particles of antimatter. But they could pass unnoticed, though, if the regions of antimatter were like small islands, quite isolated from an ocean of matter surrounding them.[4] This is precisely the case in the models of isothermal baryonic fluctuations, proposed by Alexandre Dolgov (Fig. 7.3) and Joseph Silk in "Physical Review D" in 1993, which are refer-

---

[4]One has to take into account also that there are enormous voids (empty regions) in the intergalactic space. Hence, antimatter galaxies could be very isolated from matter so that no annihilation boundaries would appear.

Fig. 7.3 Alexander Dolgov, cosmologist and astrophysicist from Novosibirsk State University (Russia) and the University of Ferrara (Italy). He is world leader in the study of possible antimatter structures in the Universe. He has been awarded the 1996 Landau-Weizmann Prize in Theoretical Physics (Israel), among other prizes. *Credit* Courtesy of the Centro de Estudos Avançados de Cosmologia (CEAC) Brasil

ence models that have given rise to many other studies and articles. In any case, the absence of this type of electromagnetic signals does give some information that allows us to estimate the matter–antimatter asymmetry. Namely, from the lack of signals one can compute some bounds on their relative abundances, although these estimates depend on the different theoretical models being used.

Likewise, no boundaries between matter and antimatter regions are observed in the CMB radiation in the form of distortions of non-thermal origin superimposed on the thermal spectrum of 2.7 K. However, in the article *A matter–antimatter universe?* by Andrew Cohen, Álvaro De Rújula and Sheldon Glashow, published in 1997 in the "Astrophysics Journal", it was shown that the boundaries of matter–antimatter annihilation in the Early Universe would produce a negligible effect on the temperature of the CMB radiation, making detection unfeasible. In a subsequent article, the first two authors further argued that not even the next generations of satellites, for the study of CMB radiation, would have the sensitivity necessary to detect the effects of a supposed matter–antimatter annihilation.

7.1.3   Antistars and Antigalaxies?

The non-observation of borders between matter and antimatter regions in the DGRB radiation, and the lack of a conclusive detection of antimatter in the cosmic rays beyond that expected to be generated in collisions and astrophysical processes, clearly indicate that primordial antimatter is a very scarce substance (if it exists at all). However, as already pointed out, it cannot be discarded that some antimatter structures exist in our observable Universe, like small islands in a vast ocean of matter. These structures might include antigas clouds, antistars, black holes made of antimatter, and antigalaxies, as predicted by the aforementioned model of isothermal baryonic fluctuations and several other subsequent proposals inspired in this model, and also predicted by a number of other independent proposals.

A crucial feature of the models based on isothermal baryonic fluctuations is that the annihilation produced at the borders between matter and antimatter regions would pull these regions away from each other. This implies that the annihilation would stop by itself and would not spread very far inside the antimatter regions. By contrast, in more traditional scenarios the annihilation at the borders between matter and antimatter regions would lower the density of energy and pressure in the annihilation "battlefield", increasing the diffusion of matter and antimatter towards each other.... until nothing is left of the antimatter regions (assuming they are the smaller ones). Let us say a few more words about the work done in the more "optimistic" antimatter-friendly scenarios.

For instance, in the article *Antimatter and Antistars in the Universe and in the Galaxy*, by Sergei Blinnikov, Alexander Dolgov and Konstantin Postnov, published in 2015 in "Physical Review D", the authors consider the possible existence of very dense stars and antistars, called baryo-dense (BD), which might have been created in the very early Universe. Those BD stars and antistars would have consisted initially of helium and antihelium, respectively, although at present most of them would be "dead", such as neutron and anti-neutron stars. They would populate the galactic halo and might add a noticeable contribution to dark matter. The authors discuss their observational manifestations, like the merging of an antistar and a star producing a huge explosion, and compute the bounds on the amounts of such antistars, showing that they are allowed to be abundant in our galaxy. They conclude that current observational data do not exclude significant amount of antimatter in our galaxy, or in other galaxies, especially in the form of the BD antistars, which move very fast and have relatively low mass.

The article *Early formed Astrophysical Objects and Cosmological Antimatter*, by Alexander Dolgov, published in 2016 in the "International Journal of Modern Physics A" brings to the attention of the reader the various inconsistencies in the current standard theory of structure formation in Cosmology and Astrophysics when faced with the astronomical observations of very old objects (at high redshifts of about $z = 10$). These observations show that the Universe at that time—some 450 million years after the Big Bang—was densely populated (Fig. 7.4) by bright galaxies, quasars, and supermassive black holes, among other objects, and contained large amounts of dust and chemical elements heavier than helium (called "metals" in astrophysics jargon). Such rich early formed structures are not expected in the standard model of structure formation, which is why there is a serious discrepancy between theory and observation. As explained by the author, "*In short, there is an avalanche of astronomical discoveries of different kind of astrophysics objects which could not be created and evolved in the available cosmological time*". Then it is shown that the model of isothermal baryonic fluctuations, which naturally leads to the creation of significant amounts of antimatter even in our galaxy, can explain all these observations at least qualitatively, since more detailed calculations will be necessary to reach a definitive conclusion.

But there are several other independent proposals to generate antimatter regions in the Universe, that we will not go into. Just let us mention the article, *An Antimatter Globular Cluster in our Galaxy: A Probe for the Origin of Matter*, by Maxim Khlopov, published in the journal "Gravitation and

Fig. 7.4 Stars or antistars? Galaxies or antigalaxies? The Hubble Ultra Deep Field is an image of a small region of space in the constellation Fornax, containing an estimated 10,000 galaxies. It was constructed from a series of observations by the Hubble Space Telescope and can look back more than 13 thousand million years (between 400 and 800 million years after the Big Bang) *Credit* Courtesy of NASA and the Space Telescope Science Institute (STScI)

Cosmology" in 1998, because of its astronomical consequences. In this work it is argued that the minimal survival scale of antimatter regions is of the order of several thousand solar masses, that is the order of the minimal mass of globular clusters. In our galaxy such regions could actually form antimatter globular clusters, which are old objects located in the galactic halo, where the interstellar gas of matter has a very low density. The continuous annihilation of the antimatter globular clusters with the gas of matter that surrounds them would take place only on the surface of the antistars and hence these annihilations would represent a very faint γ ray source.

On the observational side, there are also news concerning cosmic rays, as pointed out before. Indeed, in the recent article *Where do the AMS-02 antihelium events come from?*, by Vivian Poulin, Pierre Salati, Ilias Cholis, Marc Kamionkowski and Joseph Silk, published in "Physical Review D" in 2019, the authors discuss the origin of the Antihelium-3 and Antihelium-4 candidates (six $^3\overline{\text{He}}$ and two $^4\overline{\text{He}}$), that might have been detected by the AMS-02 experiment. They argue that common astrophysical explanations, as well as dark matter annihilations, face real difficulties in order to explain these events, and for this reason they consider the possibility that the events originate from antimatter dominated regions in the form of antigas clouds or antistars. In this respect, they present astrophysical constraints from the survival of anticlouds in our galaxy and in the early Universe, and also discuss the alternative in which the antimatter regions are dominated by antistars. Finally, they suggest that part of the unidentified sources in the *Fermi Telescope Third Source* (3FGL) *catalog* could originate from anticlouds or antistars.

Needless to say, there is always the possibility that substantial amounts of primordial antimatter exist beyond our cosmological horizon, also called particle horizon; that is, in regions of the Universe that we know nothing about because no information has ever reached us from them. Moreover, more and more known regions are moving to the other side of the horizon and will continue to do so in the future, due to the acceleration of the expansion of the Universe. As a consequence, those regions are becoming gradually invisible to us and to all other possible observers of our galaxy, as was noted in Chap. 3.

7.2 Matter–Antimatter Asymmetry in the Standard Model

Next, we will examine more closely the problem of the matter–antimatter asymmetry in the light of the Standard Model, in which there are some conserved quantities and some symmetries of great relevance to the problem at hand. Indeed, we will see that the operations or transformations denoted as C, P, CP, T and CPT are crucial to understand the symmetries underlying the processes between subatomic particles. These operations transform the processes between particles into other processes that may or may not exist. If the transformed processes exist and have analogous properties as the original ones, then one says that this operation, or transformation, is a symmetry of the laws that govern those processes. Otherwise, the transformation is not a symmetry, but if it fails to be so in only a small number of processes compared to the total, then one says that such anomalous processes violate that symmetry.

7.2.1 Conserved Quantities: q, L and B

Let us begin with the conservation of the *electric charge q*, which we have already encountered in Chap. 2. This is always conserved; therefore, the initial and final electric charges of any physical process must be the same. For this reason, charged particles are created in particle-antiparticle pairs when they are produced from very energetic photons, since these have no charge.

Lepton number L and *baryon number B* are properties that only exist for leptons and baryons, respectively, with the assignment of $+1$ for the matter particles and -1 for their antiparticles. They are also conserved quantities; that is, in the reactions between particles the sum of the lepton numbers of the initial particles must coincide with the sum of the lepton numbers of the final particles, and the same applies to the baryon number. Since baryons are hadrons consisting of three quarks, each quark has $B = 1/3$, whereas each antiquark has $B = -1/3$. In fact, lepton number conservation is the reason why in the neutron β decay: $n \rightarrow p + e^- + \overline{\nu}_e$, the emitted electron neutrino is actually an antineutrino. Namely, since baryons (like the neutron and proton) have $L = 0$ and the electron has $L = 1$, it follows that the neutral particle emitted must be an antineutrino with $L = -1$. We can also see that the baryon number is conserved in this decay because $B = 1$ for baryons, and $B = 0$ for leptons.

However, unlike the conservation of electric charge, which is exact, the conservation of the L and B numbers is not exact, at least in theory, although a process in which these numbers are not preserved has never been seen. The conservation of L and B is not exact due to a very small quantum effect, the so-called *Adler-Bell-Jackiw* anomaly, which can only be observed at much higher energies than we currently have access to.

7.2.2 Charge Conjugation C

Charge conjugation C replaces all the charges (strong, weak, q) and the numbers L and B with their opposites. The laws of electromagnetism are invariant under this transformation; hence C is a symmetry of all electromagnetic processes. Gravitation is also invariant under C because it does not even distinguish between the different charges, and the strong forces also preserve this symmetry. However, the weak interactions mediated by the bosons $W^{\pm}$ and Z^0 do violate the C symmetry. This can be deduced quite easily from any process involving the $W^{\pm}$ bosons; let us see an example.

Consider again the neutron β decay. As was explained in detail in Chap. 2, this process takes place in two phases. First, a quark d of the neutron emits a W^- boson, becoming a quark u, and then this boson decays producing one electron e^- and one electron antineutrino $\bar{\nu}_e$. But it turns out that only the matter particles with left-handed helicity, and the antimatter particles with right-handed helicity, participate in the exchanges of the $W^{\pm}$ bosons; in other words, only those particles have couplings with these bosons. Now, if we write the neutron β decay indicating explicitly the helicity of the electron and of the antineutrino we have:

$$n \rightarrow p + W^- \rightarrow p + e_L^- + \bar{\nu}_R, \qquad (7.3)$$

where L and R indicate left-handed and right-handed helicity, and we have omitted the subscript e of the electron antineutrino for clarity. The process resulting from the charge conjugation C is therefore:

$$\bar{n} \rightarrow \bar{p} + W^+ \rightarrow \bar{p} + e_L^+ + \nu_R, \qquad (7.4)$$

which is a non-existent process that does not correspond to any decay channel of the antineutron $\bar{n}$. The reason is that the W^+ boson never decays giving as products a left-handed positron e_L^+ or a right-handed neutrino ν_R, since these are particles with which it does not have any couplings. We are not even sure that the neutrino ν_R exists, as it becomes undetectable by not having interactions with the rest of the particles, except for the gravitational interactions, that are extremely feeble.

7.2.3 Parity Transformation P

The *parity transformation* P reverses all the coordinates along the three dimensions of space, which is why it is also called space inversion or reflection.

It corresponds to a mirror reflection plus a rotation of 180° about an axis perpendicular to the mirror. However, since all the known forces are invariant under rotations, the action of P effectively reduces to a mirror reflection. Again, the laws of electromagnetism and gravitation, as well as the strong forces, are symmetric under P; and again, the weak interactions violate this symmetry.

This was discovered in 1956 as a result of the famous *Wu experiment* led by the physicist Chien-Shiung Wu (Fig. 7.5), which took place in the low temperature laboratory of the US National Bureau of Standards, in Washington, D.C. This was the experiment that revealed that right-handed matter particles do not notice the weak interactions involved in the disintegrations of radioactive nuclei. The realization of this experiment was proposed by the theoretical physicists Tsung-Dao Lee and Chen-Ning Yang, as they came up with the idea that parity might not be preserved in weak interactions since this had never been tested by any experiment. This discovery earned the Nobel Prize to Lee and Yang in 1957, but it was not awarded to Wu, despite

Fig. 7.5   The Chinese-American experimental physicist Chien-Shiung Wu (1912–1997), who in 1956 designed and led the experiment in which weak interactions were found to violate the Parity symmetry. This photo was taken around the time when she received the National Medal of Science (NMS) in 1975 "for her ingenious experiments that led to new and surprising understanding of the decay of the radioactive nucleus." She was also awarded the first Wolf Prize in Physics from the Wolf Foundation in Israel. *Credit* Courtesy of the National Science Foundation, USA

the fact that she designed the experimental set up, using the radioactive isotope Cobalt-60.

The neutron β decay (7.3) transformed under P results in:

$$n \to p + W^- \to p + e_R^- + \bar{\nu}_L, \tag{7.5}$$

where it can be observed that P has exchanged the left-handed and right-handed helicities of the particles.[5] As before, this decay channel of the neutron does not exist either, as the Wu experiment showed, because the right-handed electrons e_R^- and the left-handed antineutrinos $\bar{\nu}_L$ do not have couplings with the W^- bosons. As for these antineutrinos $\bar{\nu}_L$, if they existed they would also be practically undetectable, like their antiparticles, the neutrinos ν_R.

7.2.4   CP Symmetry

The operations C and P can also be combined, that is, performed one after the other, in any order. The resulting CP transformation replaces all the charges of the particles with their opposites and also reverses their helicity. Thus, CP transforms the particles into their antiparticles, and vice versa. As a note of historical interest, we have to emphasize that it was Lev Landau, a very prominent Soviet physicist, who proposed in 1957 that the distinction between particles and antiparticles should include the inversion of the helicity and, therefore, the transformation between them is CP, not only C.

In general, the reactions between particles are symmetric under CP. That is, if a process exists, then the transformed process, by exchanging all the particles for their antiparticles, also exists and will have the same probability of being produced. The neutron β decay (7.3) is transformed under CP in the process:

$$\bar{n} \to \bar{p} + W^+ \to \bar{p} + e_R^+ + \nu_L, \tag{7.6}$$

which exists and corresponds to a decay channel of the antineutron $\bar{n}$, with the same probability as the β decay of the neutron. If this were not so, i.e. if these two decays did not occur at the same rate, then the CP transformation would not be an exact symmetry of these decay channels, but only an approximate symmetry. In that case one says that such decays violate the

---

[5]Helicity is the projection of the spin in the direction of the particle's movement. The spin remains invariant under the action of P, but the direction of the movement is reversed; consequently, P exchanges the left and right-handed helicities.

CP symmetry in a *direct* way, signaling a clear asymmetry in nature between matter and antimatter.

CP violation in K mesons.

As it turns out, the CP symmetry is violated in some very rare processes between particles, in which the weak interactions are involved (again). This has been known since 1964, the year in which James Cronin and Val Fitch (Fig. 7.6), working at the Brookhaven National Laboratory, discovered a small anomaly in the behavior of the so-called long (long-lived) kaon K_L, which is a mixture of the neutral kaon K^0 and its antiparticle $\overline{K}^0$. These particles are K mesons composed of one quark and one antiquark of types d and s; to be precise, $K^0 = (d, \overline{s})$ and $\overline{K}^0 = (\overline{d}, s)$. In addition, lacking electrical charge these mesons can mix with each other (for very technical reasons that we will not go into), so that the resulting kaons with well defined masses and lifetimes are K_L and K_S, the latter called short kaon in reference to its short lifetime compared to the long kaon K_L. In brief, the observed anomaly was the presence of a decay channel of the kaon K_L that should not exist, consisting of

Fig. 7.6 James Cronin (1931–2016) and Val Fitch (1923–2015) shared the 1980 Nobel Prize in Physics "for the discovery of violations of fundamental symmetry principles in the decay of neutral K mesons." *Credit* Courtesy of the Nobel Foundation

two pions, and was found to be due to a violation of the *CP* symmetry, which until that moment had been considered sacrosanct. This type of anomaly is referred to as *indirect violation* of the *CP* symmetry because it manifests itself through the mixing of the two neutral kaons.

This discovery had far reaching consequences because, in order to accommodate the violation of the *CP* symmetry in the Standard Model without substantially deviating from it, it was necessary to assume either the existence of a third family of elementary fermions or a second Higgs field, as proposed by Makoto Kobayashi and Toshihide Maskawa in 1973. It happened that with the two families of fermions known at that time and only one Higgs this symmetry was exact. By including one more family, or another Higgs, the Standard Model itself predicted that the *CP* symmetry had to be violated through certain parameters in its equations, as these parameters gave rise to processes occurring at different rates for particles and their antiparticles. A few years later it became evident that there was indeed a third family of fermions, and its components—the τ lepton with its neutrino ν_τ, and the quarks b (bottom) and t (top)—were gradually discovered, together with their antiparticles.

Surprisingly, it took 35 years for the first *direct violation* of *CP* symmetry to be detected, which occurs when a decay channel of a particle and the *CP*-transformed decay channel of its antiparticle, have different probabilities of being produced, as explained above. The discovery was made in 1999 and took place, almost simultaneously, in the *KTeV* experiment of the Tevatron at Fermilab and in the *NA48* experiment of the SPS (Super Proton Synchrotron) at CERN. In both experiments, processes were found in which the neutral kaon K^0 and its antiparticle $\overline{K}^0$ decay at slightly different rates, the difference being of only one part in a million.

CP violation in B mesons

Shortly afterwards, in 2001, indirect violations of the *CP* symmetry were observed also in the neutral B^0 and $\overline{B}^0$ mesons, due to similar reasons as for the neutral K^0 and $\overline{K}^0$ mesons. *B* mesons are characterized by having either a quark b or an antiquark $\overline{b}$ as one of their constituents; in particular, the B^0 and $\overline{B}^0$ mesons are composed as $B^0 = (\overline{b}, d)$ and $\overline{B}^0 = (b, \overline{d})$. Moreover, in 2004 direct violations of *CP* were also discovered in some decays of all the *B* mesons, both neutral and charged. For example, the decays

$$B^0 \to K^+\pi^- \quad \text{and} \quad B^0 \to \pi^+\pi^- \tag{7.7}$$

of the B^0 meson, are not produced with identical probabilities as the *CP*-transformed decays of the $\overline{B}^0$ meson:

$$\overline{B}^0 \rightarrow K^-\pi^+ \quad \text{and} \quad \overline{B}^0 \rightarrow \pi^-\pi^+. \tag{7.8}$$

These findings were made in the so-called *B* meson factories, which were the BaBar experiment (Fig. 7.7), at SLAC National Accelerator Laboratory in California, and the Belle experiment, at KEK in Japan. Since then, many more processes that violate the *CP* symmetry, both directly and indirectly, have been discovered. Moreover, in 2010 the LHCb experiment at CERN was added to the former *B* meson factories. The objective of this experiment is precisely to perform highly accurate measurements of the decays of the quark *b* and its antiquark $\overline{b}$, the essential components of the *B* mesons, and therefore the origin of the *CP* violations observed in these particles.

CP violation in baryons

Some *CP*-violating processes have also been observed in decays of the baryon Λ_b(*b lambda*), formed by the trio of quarks (*u*, *d*, *b*), whose mean lifetime is 1.43×10^{-12} s. Indeed, in 2017 the LHCb collaboration at

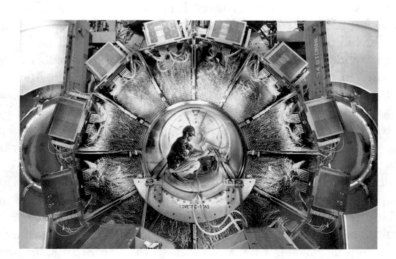

Fig. 7.7 The BaBar detector was built at SLAC to study the *B* mesons produced by the PEP-II electron–positron collider. In addition of finding direct *CP* violation for these mesons in 2004, the BaBar collaboration also achieved the first direct observation of *T* (Time reversal) violation in 2012. Although the detector was decommissioned in 2008, the analysis of the data obtained during the nine years of operation is still ongoing. *Credit* Courtesy of SLAC

CERN published in "Nature Physics" the article *Measurement of matter–anti-matter differences in beauty baryon decays*, where the researchers announced evidence of CP violation in the decay channel of the Λ_b baryon giving as by-products a proton and three pions:

$$\Lambda_b \rightarrow p \ \pi^- \pi^+ \pi^-. \tag{7.9}$$

That is, they observed differences between the probability of this decay and the probability of the *CP*-transformed decay: $\overline{\Lambda}_b \rightarrow \overline{p} \ \pi^+ \pi^- \pi^+$. This represented the first evidence for CP violation in baryons; in particular, in a "beauty" baryon.[6]

Hence, strange as it may sound, a particle and its antiparticle do not necessarily behave identically; in other words, antiparticles are not necessarily the exact mirror images of particles with respect to the "mirror" that the *CP* transformation provides.

CP violation in neutrino oscillations

CP violation is also found in the asymmetries in the oscillations of neutrinos[7] with respect to the oscillations of antineutrinos, asymmetries that have just been confirmed by the T2K collaboration after several years of experiments. Indeed, already in August 2017 the research team of the *Tokai-to-Kamioka* (T2K) experiment, which is being carried out in Japan, presented very solid indications that the oscillations of neutrinos and antineutrinos do not have identical probabilities, but until late 2019 the results were not conclusive. In a very recent article published in "Nature" in April 2020, with the title *Constraint on the matter–antimatter symmetry-violating phase in neutrino oscillations*, the researchers report a measurement observed by the T2K experiment that shows a large increase in the neutrino oscillation probability, as compared with the antineutrino oscillation probability, concluding that these results indicate *CP* violation in leptons.

Neutrino oscillations were discovered in 1998 from the detection, by the Super-Kamiokande experiment, of muon neutrinos changing into tau neutrinos. In this type of oscillations, neutrinos are only transformed into neutrinos, and antineutrinos into antineutrinos; that is, neutrinos and antineutrinos do not mix with each other. In the T2K experiment (that we already met in Chap. 6) the neutrinos are produced using an accelerator in the J-PARC laboratory of the KEK campus in Tokai; to be precise, beams

---

[6]The b quark, most commonly called "bottom", is also dubbed "beauty".
[7]Neutrino oscillations were briefly introduced In Chap. 2, Sect. 2.5.2.

of both muon neutrinos ν_μ and antineutrinos $\bar{\nu}_\mu$ are created. Then they are sent to the huge neutrino detector of the *Super-Kamioka Neutrino Detection Experiment* (Super-Kamiokande) in Kamioka, 295 km away. During their flight, a fraction of these particles oscillate and become electron neutrinos ν_e and antineutrinos $\bar{\nu}_e$, respectively; and then a very small fraction of all those neutrinos and antineutrinos, of the two kinds, interact with the water of the detector, allowing their identification. The T2K experiment analyzes the results of these oscillations and has found more electron neutrinos ν_e, and less antineutrinos $\bar{\nu}_e$, than it should if the *CP* symmetry were exact. In other words, the T2K collaboration has detected appreciable differences in the probabilities with which the muon neutrinos and antineutrinos, ν_μ and $\bar{\nu}_\mu$, oscillate becoming electron neutrinos and antineutrinos, ν_e and $\bar{\nu}_e$, respectively.

The Super-Kamiokande neutrino detector (Fig. 7.8), which is located 1000 m underground, consists basically of a cylindrical tank 39 m in diameter and 42 m in height, filled with 50,000 tons of ultrapure water. It is part of the Kamioka Observatory of the Institute for Cosmic Ray Research (ICRR) from Tokyo University and started operation in 1996. It is equipped

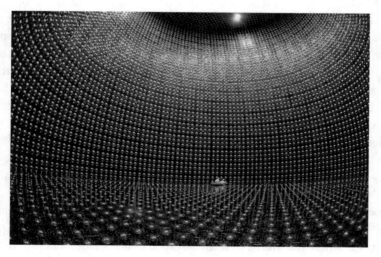

Fig. 7.8 The Super-Kamiokande neutrino detector, 1000 m underground in the Mozumi Mine (Japan), consists of a cylindrical tank 39 m in diameter and 42 m in height, filled with 50,000 tons of ultrapure water and equipped with 13,000 photo-multiplier tubes. It started operation in 1996, and in 1998 the Super-Kamiokande collaboration announced the first evidence of neutrino oscillations, from the detection of muon neutrinos changing into tau neutrinos. This was the first experimental observation supporting the hypothesis that the neutrinos have a non-zero mass. *Credit* Courtesy of Alchetron

with some 13,000 light sensors in order to register the Cherenkov radiation[8] of muons and electrons created by the interactions between the high energy neutrinos and the electrons or nuclei of the water molecules. The cones of Cherenkov light are projected as rings on the wall of the detector and are recorded by the light sensors (photomultiplier tubes).

7.2.5   Time Reversal *T*

Time reversal T consists of replacing the time parameter t in the equations describing a physical process by its opposite $(-t)$, interchanging in addition the initial and final states. The T transformation is, in fact, equivalent to record a movie of a given process and then play it backwards.

In our daily life, most events and situations are far from being symmetric under T, which is why most movies played backwards show a plethora of impossible and ridiculous features and situations that can easily be distinguished from reality. For instance, let us imagine that we record a busy street in a city on an ordinary day where nothing special is taking place. Some people will be going their way fast on the sidewalks, others will be walking the dog or hanging around at a slower pace, while a few others will be running across the street dodging the cars. Needless to say, in just seconds we would be able to distinguish if the movie is being played forwards or backwards, because in normal streets on ordinary days cars do not drive backwards, no humans or dogs walk backwards, and even less someone would try to cross the street in between the cars running backwards.

But the equations describing the fundamental laws of Physics are pretty much symmetric under the exchange of t by $(-t)$. For example, let us consider a movie made in a telescope recording the movement of a planetary system around its central star. The planets, if they were formed together at roughly the same time, will be all rotating in the same clockwise or counter-clockwise direction. In fact, the laws of gravitation and celestial mechanics allow both senses of rotation of planets around the stars, and the actual sense in a given system is solely determined by the conditions that were present at the time when the stellar system began to take shape. These two senses of rotation can be swapped with each other by the action of T, obviously, but also by the action of P (as can be easily seen by watching the hands of a clock which is placed in front of a mirror). Consequently, judging from the images of the planets revolving around their star there is no way to distinguish

---

[8]Charged particles moving faster than photons in water or other material medium give rise to a cone of light known as Cherenkov radiation. Its cause is similar to that of a sonic boom when an object travels in the atmosphere faster than the speed of sound.

whether the movie is played forwards or backwards,[9] neither if it was made by recording the real images of the planets or by recording their mirrored images produced by the internal mechanism of the telescope.

Specially interesting are the situations for which the T and C transformations produce the same effects. To see this let us record the trajectory of a charged particle in an electromagnetic field, very much like taking photographs of particles tracks in the cloud chambers of Chap. 4. If we run the movie backwards what we will see is exactly the trajectory of another almost identical particle, but with the opposite electric charge than the original. So, charged particles and their antiparticles do follow the same trajectories in the presence of electromagnetic fields, but they move in opposite directions, as if the antiparticles were identical to the particles but *traveling backwards in time*, or conversely. Such a simple fact, nevertheless, has led to enormous confusion and misconceptions as to whether time travel is possible and whether positrons are electrons coming from the future traveling backwards in time. Actually, the situation between particles and antiparticles is completely symmetrical in this respect, as just said, and therefore the matter particles could be regarded as being antimatter particles traveling backwards in time instead of the other way around. In conclusion, neither particles nor antiparticles travel backwards in time, they simply travel in opposite directions in the presence of electromagnetic fields.

In Particle Physics, time reversal is an almost exact symmetry; that is, the great majority of reactions between particles are symmetric under T, as they are under CP. In fact, as far as we know, all the processes that violate the CP symmetry also violate the T symmetry, and vice versa. There are good reasons for this, as we will see below. In any case, the first direct observation of T violation was achieved in 2012 by the BaBar collaboration at SLAC (see Fig. 7.7). It was detected in the behavior of B^0 mesons. The BaBar Collaboration consists of approximately 600 physicists and engineers from 75 institutions in 13 countries. Although the detector was shut down in 2008, data collected during the nine years of operation are still under analysis.

7.2.6 *CPT* Symmetry

The operations C, P and T can be combined in any order and the resulting CPT transformation is believed to be an exact symmetry of all processes

---

[9]Long recordings of the planetary system would allow us to distinguish between forwards or backwards, however, since the orbits of planets and satellites change in a predictable (although complicated) way due to several effects. At present, the Earth is slowly moving away from the Sun at about 15 cm per year and the Moon is moving away from the Earth at about 38 mm per year.

between particles. Indeed, CPT violation has never been observed. The CPT transformation is actually a symmetry of the Quantum Field Theory on which the Standard Model is built. As a result, if the CPT symmetry were violated, the Standard Model would not stand on solid ground.

The CPT symmetry implies that particles and their antiparticles must have the same mass, the same spin, the same magnetic moment, and the same mean lifetime. Regarding the latter, one has to take into account that particles decay through more than one channel and the sum of all the contributions of the different channels is what determines their mean lifetime. Only this must be identical for particles and their antiparticles, but not the contributions of any specific decay channel for the particle and the corresponding CP-transformed channel for the antiparticle. As to the magnetic moment—the sensitivity of a particle to magnetic fields—there are precision measurements for protons and antiprotons performed at CERN's Antimatter Factory since 2013, finding no difference between them, as will be discussed in next chapter.

Now, if the CPT symmetry is always preserved, then the CP violations must be compensated by simultaneous violations of the T symmetry. This is the indirect way to detect T violation and has been the only way for nearly 50 years since the discovery of CP violation in 1964; to be precise, until 2012 when the BaBar collaboration made the first direct observation of T violation, as noted above.

7.2.7 Particle-Antiparticle Oscillations of Neutral Mesons

A very curious peculiarity, both of the neutral K mesons (K^0 and $\overline{K}^0$) and of the neutral B mesons (B^0, $\overline{B}^0$, B_S^0 and $\overline{B}_S^0$), is that they change their personality at a high rate, oscillating in a dizzying way between their particle and antiparticle identities. This property, perfectly explained within the Standard Model, is simply due to the exchange of two $W^\pm$ bosons between the quark and the antiquark that form the meson, for each oscillation. Hence, it is the weak interactions that produce this interesting phenomenon, which is why sometimes they also produce CP violations that are reflected in those oscillations.

Take, for instance, the case of the B^0 meson and its antiparticle $\overline{B}^0$, constituted of the form: $B^0 = (\overline{b}, d)$ and $\overline{B}^0 = (b, \overline{d})$. As seen in the Feynman diagram shown in Fig. 7.9, that represents the time evolution of these particles, the $\overline{B}^0$ meson is transformed into B^0 through the exchange of two $W^\pm$

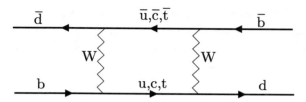

Fig. 7.9 Particle-antiparticle oscillations of the neutral B^0 and $\overline{B}^0$ mesons

bosons between the quark-antiquark pair $(b, \overline{d})$ that composes it. In the first exchange the quark b becomes another quark, which may be u, c, or t; while the antiquark $\overline{d}$ becomes another antiquark, which may be $\overline{u}$, $\overline{c}$ or $\overline{t}$. We do not put arrows indicating which particle emits and which absorbs the W boson, as both possibilities can occur, and depending on which is the particle that emits in each case, then the boson exchanged will be the W^+, or else the W^-. This is easy to see simply by applying electric charge conservation at each intersection of three particles, or vertex.

The rate of these oscillations is really shocking. In 2006, it could be determined that the fastest oscillations, which take place between the meson $B_S^0 = (\overline{b}, s)$ and its antiparticle $\overline{B}_S^0 = (b, \overline{s})$, occur at a rate of 2.8×10^{12} per second! This was achieved in the CDF detector of Fermilab's Tevatron, the same detector in which the top quark was discovered in 1995. Nevertheless, the mean lifetime of these particles is very short, $\tau_B = 1.5 \times 10^{-12}$ s, hence there is hardly any time for many of these oscillations to occur.

7.3 Baryogenesis

The origin of the problem of the matter–antimatter asymmetry in the Universe lies, in part, in the general consensus among physicists that matter and antimatter were created in identical quantities less than a microsecond after the Big Bang. Obviously, we have no evidence that this idea is correct, but it is consistent and in line with the knowledge we have, both from Particle Physics and Cosmology. Moreover, as was noted in Chap. 3, the majority of cosmologists believe that there was a stage of extremely rapid expansion of the Universe, called cosmological inflation, which would have diluted the initial conditions regardless of what they were. This would imply that, in practice, matter and antimatter were created in identical quantities, even though this may not be exactly the case. In spite of this, as we have seen, the non-observation of gamma radiation boundaries between regions of matter and regions of antimatter, leads us to conclude that primordial antimatter, if it exists at all in the observable Universe, must be present in very small quantities compared to ordinary matter, quantities that the AMS experiment is trying to determine from its privileged position at the International Space Station.

7.3.1 The Great Annihilation

Now let us return to the Big Bang and to the events that followed, in search of an explanation of what caused the antimatter to disappear. These events are deduced quite accurately from our knowledge of Particle Physics and Cosmology, since the particle collisions in the large accelerators reproduce, approximately, the particle collisions in the Universe in the first moments after the Big Bang. To start, we find that if the matter particles were created in identical quantities as their antiparticles, in the first trillionth of a second (10^{-12} s), they had to annihilate each other almost completely immediately afterwards. This surprising result is due to the prevailing conditions at that time: very high density and temperature, together with very little space available for the particles and antiparticles to be able to separate and avoid mutual extinction.

This somewhat gloomy scenario also implies that the problem of matter–antimatter asymmetry is not so much why the antimatter disappeared, but rather why ordinary matter did not disappear with it; that is, how it was capable of surviving the *Great Annihilation* between matter and antimatter, which occurred less than a second after the beginning of the Universe. Let us have a closer look at this.

In the first instants, the particle-antiparticle pairs that were created, were also annihilated almost at once, producing very energetic photons that, in turn, used their energy to create new particle-antiparticle pairs. But before they were annihilated, those particles - quarks, antiquarks, electrons, positrons, muons, etc.—had time to interact with some others through the exchange of the mediating bosons: gluons, photons, $W^{\pm}$ and Z^0. Therefore, the possibility exists that these interactions created a tiny excess of quarks over antiquarks and leptons over antileptons.

Meanwhile, our minuscule Universe was expanding rapidly, cooling down, which translated into a decrease in the energies with which the collisions occurred. Once the temperature dropped enough, the photons no longer had the necessary energy to create new particle-antiparticle pairs, while the still existing ones continued annihilating each other. Then, the quarks and antiquarks started joining each other to form hadrons: baryons, antibaryons and mesons. The vast majority of these hadrons spontaneously disintegrated in less than a microsecond after they were created, due to their unstable nature, in particular all the mesons. And almost all the baryons and antibaryons decayed too, all except the protons, the neutrons and their antiparticles. But these "surviving" baryons, which remained because they were long-lived, were not much better off since the immense majority of them disappeared as well when the quarks of protons and neutrons were annihilated by the antiquarks of both antiprotons and antineutrons in their environment.

The balance, a second after the Big Bang, was devastating for the baryons because, as a result of the Great Annihilation, only one was left for about thousand million photons. To be precise, there were about 1.61×10^9 photons for each baryon left, as can be deduced from the estimates of the baryon abundance in the Universe today and from the CMB radiation. As for the antibaryons, if some antiprotons and antineutrons survived, there had to be in negligible amounts, such as one antibaryon among millions of baryons, at most.

Nevertheless, something very interesting happened as a result of the Great Annihilation: the neutrinos and antineutrinos, which had survived in large quantities because they barely interact, decoupled from the other particles. Lacking electric charge, they were already decoupled from the photons, but now they also decoupled from the baryons due to their scarcity. Consequently, the Universe became transparent to the primordial neutrinos and antineutrinos between one and two seconds after its beginning. This implies that there must necessarily exist a cosmic background radiation of these particles, with a very low energy, which we are not able to detect with our current instruments, as was pointed out in Chap. 2.

Some 10 seconds after the Big Bang, as the Universe expanded and cooled, the protons and neutrons began to bond together to form atomic nuclei. This process lasted 20 minutes, approximately, and only very light nuclei were created; namely, the nuclei of deuterium ^{2}H, tritium ^{3}H, helium ^{3}He and ^{4}He, lithium ^{7}Li and beryllium ^{7}Be. The tritium and beryllium nuclei, however, being unstable disintegrated ending up as ^{3}He and ^{7}Li nuclei instead. The "low" temperatures (hundreds of millions of degrees!) did not allow more nuclear reactions, and the result of this *Primordial Nucleosynthesis* was essentially 8% of ^{4}He nuclei surrounded by 92% of protons plus some trace amounts of the other isotopes. These were about three ^{2}H nuclei and one ^{3}He nucleus per 100,000 protons and about four ^{7}Li nuclei per 10,000 million protons.

The heavier nuclei up to the iron isotope ^{56}Fe were only created much later within the stars, while the even heavier nuclei had to be produced in very energetic astrophysical processes, such as supernova explosions and neutron stars mergers. However, it still took 380,000 years for the light nuclei to be able to capture electrons and form atoms, which released the thousands of millions of photons per baryon that make up the CMB background radiation.

7.3.2 Baryogenesis and the Sakharov Conditions

The scenario of the Great Annihilation and Primordial Nucleosynthesis just described is accepted by the vast majority of scientists today, and the creation of the surplus of baryons over antibaryons in the first moments after the Big Bang is called *baryogenesis*. It should be noted that this small surplus is what made it possible for the baryons to survive and give rise to our material Universe. This excess of baryons over antibaryons should not be confused with the parameter η of the *baryon asymmetry*, which is the ratio between this excess—the number of baryons N_B minus the number of antibaryons $N_{\bar{B}}$—and the number of photons N_γ in the CMB radiation, and has the approximate value:

$$\eta = \frac{N_B - N_{\bar{B}}}{N_\gamma} \approx 6.19 \times 10^{-10}. \tag{7.10}$$

This means that the Universe has about 1.61×10^9 photons for each baryon, as mentioned above.

At present, the mechanism that produced the amount of baryogenesis necessary to create the matter of our Universe is unknown. Within the Standard Model, only mechanisms that give rise to much smaller quantities are known, about a factor of 10^{12} smaller, or even less, although many proposals have been made extending this model in different manners.

One of the first scientists to deal with this problem in depth was the Russian dissident Andrei Sakharov (Fig. 7.10), a nuclear physicist who contributed to the construction of the Soviet hydrogen bomb before devoting himself to Particle Physics and Cosmology. On January 1st, 1967 he published the article *Violation of CP invariance, C asymmetry, and Baryon asymmetry of the Universe*, in the Soviet journal "JETP". The essence of that article is often rephrased as the proposal of three conditions, to be met by any theory describing elementary particles in order to generate an excess of baryons in an expanding Universe. These conditions, named after the author, are still considered the basic ingredients of baryogenesis and are as follows:

Fig. 7.10 Andrei Sakharov (1921–1989) was a nuclear physicist who contributed to the construction of the Soviet hydrogen bomb before switching to Particle Physics and Cosmology. He was an advocate of disarmament and human rights in the Soviet Union and became known all over the world for his activities as Russian dissident, for which he received the Nobel Peace Prize in 1975. The Sakharov Prize, awarded annually by the European Parliament for human rights activists, is named in his honor. *Credit* Courtesy of the Atomic Archive

The Sakharov conditions

1. First of all, there must be processes, or interactions between the particles, that do not preserve the baryon number; in other words, that violate the conservation of B. This condition, although obvious, is however not sufficient for baryogenesis to occur because there might be other reactions that could reverse this effect.

2. Second, there must be processes that violate the symmetries C and CP; i.e. which are not symmetric under the charge conjugation C nor under the CP operation that transforms particles and antiparticles into each other. This condition also seems obvious, since it introduces an asymmetry between particles and antiparticles that may result in an excess of the former over the latter in the first instants of the Universe.

3. Third, the processes which do not preserve the baryon number B, or which violate the symmetries C or CP, must be performed out of thermal equilibrium. This means that those processes are not reversible, since thermal equilibrium can be considered almost as a synonym of reversibility

in the reactions between particles. In an expanding Universe this is satisfied when the rates at which the interactions take place are smaller than the rate of the Universe's expansion.

Let us now review what the Standard Model says about the first two Sakharov conditions. As was explained in Sect. 7.2, in this model there are processes that violate the conservation of the baryon number B, as well as processes that violate the symmetries C and CP. The question is then if these processes could produce the required amount of baryogenesis that would explain the matter–antimatter asymmetry in our Universe.

In the first place, in 1976 Gerard 't Hooft (Fig. 7.11) showed that the conservation of the baryonic number B is violated through the Adler-Bell-Jackiw quantum anomaly, as mentioned in Sect. 7.2.1. Moreover, an important role is played by certain field configurations called *sphalerons*, discovered by Frans Klinkhamer and Nicholas Manton in 1984, which transform three antileptons (positrons, antineutrinos, etc.) into three baryons. Hence, sphalerons violate the conservation of the baryon number B, as well as the conservation of the lepton number L, although the number B-L is preserved. Nevertheless, these processes require very high energies, which is why they have never been observed, but they could have contributed in a crucial way to the creation of baryons in the first instants of the Universe, due to the very high densities and temperatures that prevailed at that time.

Regarding the violation of the symmetries C and CP, we have seen that the weak interactions strongly violate the C and P symmetries whereas they violate very slightly the CP symmetry. However, no violations of the CP symmetry have been observed involving the constituents of atoms. They have only been observed for the B and K mesons, the Λ_b baryons, and the three species of neutrinos (via neutrino oscillations). Furthermore, the contribution to baryogenesis of the known Standard Model processes which violate CP has been calculated,[10] and the result is that the amount of matter generated would be about 10^{12} times smaller than observed.

The third Sakharov condition, the presence of processes out of thermal equilibrium, i.e. non-reversible processes, depends on the details of the specific cosmological model at hand, not on the Standard Model itself. In any case, although the processes between elementary particles are all reversible in principle, in an expanding Universe one expects non-reversible processes

---

[10] In 1994 Belen Gavela, then at CERN, Pilar Hernández, then at Harvard University, and two other physicists were the first who proved that the baryon asymmetry generated by the processes that violate the CP symmetry in the Standard Model is many orders of magnitude smaller than the observed one, at least 10^{12} times smaller.

Fig. 7.11 Gerard 't Hooft, from Utrecht University (The Netherlands), discovered in 1976 that the baryon and lepton numbers, B and L, were not exactly conserved in the Standard Model due to the Adler-Bell-Jackiw anomaly. He was awarded the 1999 Nobel Prize in Physics, along with Martinus Veltman, for his crucial contributions to put the quantum field theory of the Standard Model on solid ground, as a mathematically consistent theory. Besides Particle Physics, he has always been very active in theoretical research on black holes, Quantum Gravity and foundations of Quantum Mechanics. *Credit* Author: Alex Kok. Courtesy of Gerard 't Hooft

as well. The reason is two-fold. First, as the Universe expands and cools down, some reversible processes cease to be so when they reach certain temperatures characteristic of those processes. A good example is the Great Annihilation between particles and antiparticles, which took place in our Universe in several stages corresponding to the temperatures at which the photons no longer had enough energy to produce the particle-antiparticle pairs again. These temperatures were slightly different for each type of particle-antiparticle pair because their masses were also different, and therefore different energies were required to create them. Second, one also has to take into account the influence of the expansion of the space itself. In this context, a reaction between particles is in thermal equilibrium if the rate of interaction between its components is faster than the rate of expansion of the Universe. If this is not the case, the products of a given reaction may move so far apart that they cannot interact again to produce the inverse reaction.

Hence, we have seen that the Standard Model satisfies the first two conditions of Sakharov to produce baryogenesis, but only in a qualitative way. In quantitative terms, the violation of the *CP* symmetry in this model is very insufficient to account for the amount of matter observed in our Universe in comparison with the amount of photons of the CMB radiation, whose ratio is given in terms of the baryon asymmetry parameter η (7.10). Therefore, it is necessary to look for theories and mechanisms beyond the Standard Model to explain the observed baryon asymmetry.

Indeed, many ideas have been proposed over the years, including supersymmetry and Grand Unified Theories (GUTs), in which the strong, weak and electromagnetic forces, all have a common origin at very high energies that makes them one and the same force. The case of GUTs is especially interesting because they have an additional source of non-conservation of the *B* and *L* numbers through new interactions mediated by new gauge bosons, opening new avenues for baryogenesis in this manner. Anyway, among the many proposals that have been put forward to explain baryogenesis, there is a clear favorite, for about 35 years now, that might solve this problem via lepton generation. Let us have a closer look at this proposal.

7.4 Leptogenesis

In 1986 Masataka Fukugita, from Kyoto University, and Tsutomu Yanagida, from Tohoku University, in Japan, published the article *Baryogenesis without Grand Unification* in "Physics Letters B". As indicated by the title, in this article the authors pointed out a mechanism to generate cosmological baryon number excess without resorting to GUTs, that were prevalent at that time. This mechanism is a fairly simple modification of the Standard Model, which consists of extending it by adding hyper massive neutrinos of Majorana type, with masses of the order of about a thousand million GeV. These particles are hypothetical elementary fermions with no electric charge, which is why they are called neutrinos, and their behavior is described by an equation proposed by Ettore Majorana in 1937, not by the Dirac equation of 1928 which describes the behavior of "almost" all known elementary fermions (see below).

An important virtue of these particles - if they actually exist - is that their interaction with the ordinary neutrinos would give to the latter their tiny masses in a very natural way, through the so-called *seesaw* mechanism. But their most crucial property is that their decay at the beginning of the Universe would create as by-products leptons and antileptons; in particular, ordinary neutrinos and antineutrinos, in different numbers, i.e. in an asymmetric manner, thereby violating the conservation of the lepton number L. This is due to the fact that the Majorana neutrinos are their own antiparticles and they decay into ordinary leptons and antileptons sort of randomly, without any special preference between them (Fig. 7.12).

The generation of an excess of lepton number L is referred to as *leptogenesis*. This would then translate into the generation of an excess of baryon number B through the sphalerons, which would convert three antileptons into three baryons, transmuting part of the leptogenesis into baryogenesis and violating the conservation of the baryonic number B, as a result. In addition, the decay of the hyper massive neutrinos would occur out of thermal equilibrium once the temperature of the Universe descended below the energy associated to their masses. Moreover, this scenario also provides sufficient amounts of C and CP violation for the actual baryogenesis to take place. Hence this modification of the Standard Model, adding hyper massive Majorana neutrinos, satisfies the three Sakharov conditions and would also produce the amount of baryogenesis required to explain the quantities of matter that we observe in the Universe.

To the original model of Fukugita and Yanagida many variations have been added, as expected, including some supersymmetric versions, but their essence is the same. For example, it has been found that successful leptogenesis is also possible by means of much lighter neutrinos, called *sterile neutrinos*, which might be detected at the LHC at CERN and by the experiments Belle II and T2K in Japan. The sterile neutrinos could account, in addition, for

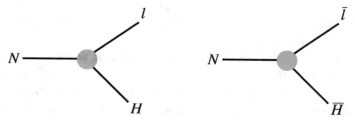

Fig. 7.12 The hyper massive Majorana neutrino N would decay in the first instants after the Big Bang, before the electroweak phase transition took place, when the Universe was extremely hot and dense. It would decay either via a lepton and a Higgs particle (left) or via an antilepton and another Higgs particle (right). Prior to the electroweak phase transition, the Higgs field had several components, not just the unique one that remained afterwards and surfaced when the Higgs boson was discovered in 2012 at CERN

neutrino oscillations and for most of the dark matter, as in a model proposed in 2005 by Takehiko Asaka and Mikhail Shaposhnikov.

It should also be noted that the model of Fukugita and Yanagida was not the first attempt to produce baryogenesis via leptogenesis, but was the first model that worked out well, although the other research lines have not been completely abandoned and have been improved over the years. For instance, leptogenesis via GUTs might be directly probed by primordial gravitational waves. Also, there is the possibility of leptogenesis in which the role of the very heavy Majorana neutrinos is played by the so-called *inflaton*, which is assumed to be the particle—and the corresponding quantum field—responsible for the primordial inflation of the Universe. This scenario would most likely leave an imprint also on the primordial gravitational waves.

7.5   Ettore Majorana and His Fermions

We will end this chapter with a few more comments about Ettore Majorana (Fig. 7.13), the author of the very relevant equation that describes a type of elementary fermions, with no electric charge, which are antiparticles of themselves. The proposal was made in the article *Teoria Simmetrica dell'Elettrone e del Positrone* (A Symmetric Theory of Electrons and Positrons), published in the Italian journal "Il Nuovo Cimento" in 1937. It must be stressed that these fermions could actually exist; in fact, the known neutrinos of the Standard Model could be such fermions, since it is not known whether they satisfy the Majorana equation, or they are subject to the Dirac equation, like all other known elementary fermions (quarks and charged leptons, along with their antiparticles). Moreover, a generic prediction of leptogenesis is that the known neutrinos of the Standard Model are also Majorana fermions because the neutrinos which acquire mass through the seesaw mechanism are such. If this prediction were true, it would help solving long-lasting problems in Particle Physics and Cosmology.

On 26 March 1938, this Italian physicist, which was part of the Enrico Fermi group in Rome, known as "the Via Panisperna boys", mysteriously disappeared—supposedly during a boat trip from Palermo to Naples (Italy)—when he was only 31 years old. However, judging by the circumstances that surrounded his disappearance, it seems that it was voluntary and premeditated, and it is not even clear that he stepped on that boat since he was traveling alone. The hypothesis shared by his family and others who knew him well, in particular by his confessor Monsignor Riccieri (he was a devout Catholic), was that Majorana retreated to a nearby monastery. Apparently,

Fig. 7.13 Ettore Majorana (1906–?) was an extremely creative and gifted physicist, who joined the group of Enrico Fermi in Rome. In 1937 he proposed the existence of a type of neutral fermions that are antiparticles of themselves, which later became known as Majorana fermions or Majorana neutrinos. He disappeared in 1938, supposedly during a boat trip from Palermo to Naples, what gave rise to many hypothesis, including suicide, but his family believed he had retreated to a nearby monastery. In 2006, the Majorana Prize was established in his memory. *Credit* Mondadori Publishers. Courtesy of Wikimedia Commons

he was going through a major personal crisis for the last four years, heading towards mysticism (essentially, he had become a "hermit"). Presumably, a six-month visit he paid to Heisenberg in Nazi Germany, in 1933, could have had some influence in that crisis, because when he returned to Rome he was sick in more than one sense, including acute gastritis and nervous exhaustion.

When Majorana vanished, Enrico Fermi (Fig. 6.9, in Chap. 6) told to his wife: "*Ettore is too intelligent. If he has decided to disappear, no-one will be able to find him*". While still in Rome, just before emigrating to the USA in 1938, Fermi also declared:

There are several categories of scientists in the world; those of second or third rank do their best but never get very far. Then there is the first rank, those who make important discoveries, fundamental to scientific progress. But then there are the geniuses, like Galilei and Newton. Majorana was one of these.

Indeed, although he published less than ten papers,[11] his legacy was most relevant since the Majorana equation and the Majorana fermions were, and still are nowadays, at the forefront in several areas of scientific research, especially Particle Physics, Mathematical Physics and Condensed Matter Physics.[12] In 1962, the Ettore Majorana Foundation and Centre for Scientific Culture was established in Erice, Sicily (Italy). This organization sponsors the International School of Subnuclear Physics, where every summer since 1963 scientific world leaders, the authors of relevant discoveries or contributions, are invited to teach students from all over the world.

In 2006, the conference "Ettore Majorana's legacy and the Physics of the XXI century" took place at the University of Catania, the city where he was born in 1906, to celebrate the centennial of his birth. In addition, the Majorana Prize was established in his memory. For that occasion, Antonino Zichichi, a prominent Italian physicist and director of the Center for Scientific Culture in Erice, wrote a very interesting article, *Ettore Majorana: genius and mystery*, in the CERN magazine "CERN Courier". Some of the most interesting anecdotes recounted by Zichichi are reproduced here, as follows.

In 1962, many illustrious physicists participated in a ceremony where the Erice School was dedicated to Ettore Majorana. In that occasion, Robert Oppenheimer (Fig. 4.2, in Chap. 4) told to Zichichi the following episode from the time of the Manhattan Project:

> There were three critical turning points during the project, and during the executive meeting to address the first of these crises, Fermi turned to Eugene Wigner and said: "*If only Ettore were here.*" The project seemed to have reached a dead-end in the second crisis, during which Fermi exclaimed once more: "*This calls for Ettore!*" Other than Oppenheimer (the project director), three people were in attendance at these meetings: two scientists (Fermi and Wigner) and one military general. After the top secret meeting, the general asked Wigner who this Ettore was, and he replied: Majorana. The general asked

---

[11] He left the work of the last four years essentially unpublished. This work was finally published in 2009, in English, in the Springer book *Ettore Majorana: Unpublished research notes on theoretical physics,* by S. Esposito, E. Recami and A. van der Merwe.

[12] In general, in Condensed Matter Physics, Majorana bound states can appear as quasiparticles and collective excitations—the collective movement of several individual particles—not as single or elementary particles. Configurations analogous to Majorana particles were found in 2012 in laboratory experiments on hybrid semiconductor-superconductor wire devices. Majorana fermions may also emerge as quasiparticles in quantum spin liquids, and were observed in 2016. Majorana bound states can also be realized in quantum error correcting codes, which could be used to process quantum information in quantum computers.

where Majorana was so that he could try to bring him to America. Wigner replied: *"Unfortunately, he disappeared many years ago"*.

In 1932 Frédérick Joliot and Irène Curie discovered a neutral particle that could expel a proton from an atom. Their interpretation was that it must be a very energetic photon, because at the time it was the only known particle with no electric charge. Majorana immediately realized that it must be a particle about as massive as the proton:

> Majorana explained to Fermi that the particle discovered by Joliot and Curie had to be as heavy as a proton, while being electrically neutral, thus a fourth particle had to exist, a proton with no charge. Fermi told Majorana to publish his interpretation of the French discovery right away. Majorana, however, did not bother to do so.

The discovery of the neutron was made just two months later by James Chadwick. Bruno Pontecorvo, who predicted the existence of neutrino oscillations many years before their discovery, recalled the origin of Majorana neutrinos in the following way:

> Dirac discovers his famous equation describing the evolution of the electron; Majorana goes to Fermi to point out a fundamental detail: *"I have found a representation where all Dirac γ matrices are real. In this representation it is possible to have a real spinor that describes a particle identical to its antiparticle"*. *"Brilliant"*, said Fermi, *"Write it up and publish it"*. Remembering what happened with the neutron discovery, Fermi wrote the article himself and submitted the work under Majorana's name to the scientific journal Il Nuovo Cimento. Without Fermi's initiative, we would know nothing about the Majorana spinors and Majorana neutrinos.

In March 2011, most surprisingly, the Attorney's Office in Rome announced in the media an inquiry regarding the statement made by a witness about meeting with Majorana in Buenos Aires (Argentina) in the 1950s. And in February 2015, the same Office released another statement, based on new evidence, declaring that Majorana was alive between 1955 and 1959 in Venezuela. Afterwards, the Attorney's Office declared the case officially closed, having found no criminal evidence related to the disappearance of Ettore Majorana.

8

Experiments with Antiatoms

8.1 Preliminaries

In Chap. 4 we saw that in 1995 a team of CERN scientists succeeded in producing the first antimatter atoms. They were the simplest ones: antihydrogen atoms $\overline{H}$, consisting of a single antiproton $\overline{p}$ in the nucleus and a positron e^+ in the shell. This happened exactly 40 years after the antiproton was created in the Bevatron at Berkeley, and with it a new era was opened in the study of antimatter.

However, the first antihydrogen atoms, created in CERN laboratories and later also in Fermilab, moved very fast, with relativistic speeds close to the speed limit c, so they could not be trapped to be analyzed and immediately collided with the surrounding matter, being annihilated as a result. This situation changed drastically in the year 2000 when the Antiproton Decelerator AD came into operation at CERN (Fig. 8.1), which, using two successive cooling techniques in addition to appropriate electric fields, was able to reduce the speed of antiprotons substantially, up to a tenth of c. In this way, when these "slow" antiprotons were mixed with positrons in the various experiments that were carried out, atoms of $\overline{H}$ were formed that did not move as fast as their predecessors and could therefore be confined for a long enough time to be analyzed.

Five experiments were conducted simultaneously in the enormous hall of the AD decelerator—the AD hall—with the aim of studying in detail several aspects of antimatter. They combined two completely different, but complementary, leading technologies in order to produce and investigate antiprotons and antiatoms. First of all, high energy particle accelerators in

© Springer Nature Switzerland AG 2021
B. Gato-Rivera, *Antimatter*,
https://doi.org/10.1007/978-3-030-67791-6_8

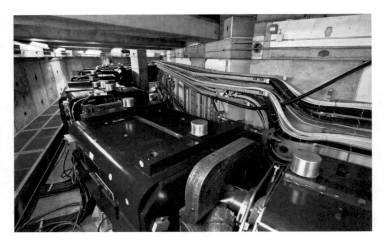

Fig. 8.1 Section of the Antiproton Decelerator AD, 182 m long, at CERN. *Credit* Courtesy of CERN

order to obtain the antiproton beams, and then low energy Atomic Physics procedures making ample use of electromagnetic traps and lasers, in order to catch, manipulate and study the antiprotons and the antiatoms thus obtained.

It should be noted that *Penning traps*, which are small electromagnetic cages, are an essential tool used by all the experiments installed in the AD hall (Fig. 8.2). They were invented in 1973 by Hans Dehmelt,[1] who was inspired by the functioning of another device made by Frans Penning in the 1930s. They allow to perform with great accuracy measurements of some properties of a single ion or a single charged particle, like an electron, a proton, a positron or an antiproton, although their main purpose in most CERN's experiments is to capture large quantities of antiprotons and positrons and bring them into contact for the production of antihydrogen atoms.

These experiments are currently shut down together with the entire accelerator complex at CERN for a major upgrade, as explained in Chap. 6, but will be resumed in August 2021. The research teams try to determine whether antimatter behaves exactly like matter or, on the contrary, some differences exist. This is particularly important because, as explained in the previous chapter, it is still not fully understood why antimatter disappeared from our Universe in its practical totality, annihilating itself against matter, whereas the

---

[1]The first "guest" trapped by Dehmelt in his newly invented device was a single electron, whose magnetic properties he measured. In 1984 he was able to store a single positron for three months in a trap half the size of a human thumb; by measuring its magnetic properties he proved that electrons and positrons are electric and magnetic mirror images of one another. In 1989 Dehmelt was awarded the Nobel Prize for developing his electromagnetic trap, that he named after Frans Penning.

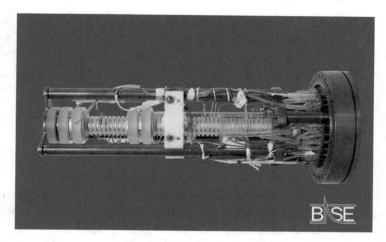

Fig. 8.2 Penning trap system of the BASE Collaboration. It consists of four Penning traps: a double Penning trap for precision frequency measurements, a cooling trap for efficient single particle cooling and a reservoir trap. This unique device—invented by BASE—traps a cloud of antiprotons from the AD decelerator and stores them for an arbitrary long time. *Credit* Stefan Sellner, Fundamental Symmetries Laboratory, RIKEN, Japan

latter survived in fairly big amounts in comparison with the former. Consequently, any difference found between the properties of particles, or atoms, of matter and the properties of their antimatter counterparts could help us solve the enigma of the disappearance of primordial antimatter. These measurements provide, in addition, stringent tests of CPT invariance, which is the most fundamental symmetry in the Standard Model of Particle Physics, as was pointed out in Chap. 7.

The antiprotons at CERN are produced in the proton accelerator PS (Proton Synchrotron), by colliding a proton beam of about 25 GeV of energy onto a metal block. Among the many resulting by-products, there are also numerous antiprotons, which are separated and directed to the AD decelerator using appropriate filters and devices. Let us remember that one of the possible processes to produce antiprotons in this manner is $p + p \rightarrow p + \overline{p} + p + p$. That is, a proton of the beam hits a proton of an atomic nucleus of the metal and the energy of the collision is inverted in creating a proton-antiproton pair, while the initial protons "reappear" due to the hadronization magic resulting from strong interactions between quarks, antiquarks and gluons.

The functioning of the AD decelerator is quite similar to that of the circular accelerators. Basically, it is a ring composed of magnets that bend the trajectories of the antiprotons and confine them in thin beams, while very intense electric fields slow them down. In addition, consecutive cooling is applied to help keeping the antiprotons on track. These spend some time in the decelerator ring, reducing their speed until their energy reaches 5.3 MeV, which is the minimum that can be achieved with this machine; and finally, the antiprotons are supplied to the different experiments that are performed in the AD facility.

8.2   The Experiments

Next, we will make a brief description of the five experiments that were taking place in the AD hall before they were discontinued on November 13, 2018. They have the names: ATRAP, ALPHA, ASACUSA, BASE and AEGIS. As we will see, an essential tool used by three of these experiments is atomic spectroscopy, in particular applied to the hydrogen atom, H. For this reason, in Appendix A we present some basic notions about this subject. Furthermore, here we will also give some details about the ATHENA experiment—the predecessor of the ALPHA experiment—and about the new experiment GBAR and the new antiproton decelerator ELENA. Figure 8.3 shows the current layout of all the experiments in the AD hall. Finally, we will also devote some lines to the FAIR complex, currently under construction in Darmstadt (Germany), which also plans to carry out experiments with antiprotons.

8.2.1   ATHENA and ATRAP

The first experiment that successfully produced a considerable amount of antihydrogen atoms $\overline{H}$, thanks to the services of the antiproton decelerator AD, was the ATHENA experiment. It created around 50,000 antiatoms in August 2002, and a few weeks later the ATRAP experiment also succeeded in producing remarkable quantities of these. In both experiments, the antiprotons coming from the AD decelerator had to be slowed down much more before putting them in contact with the positrons, by employing various techniques that used electric fields to slow them down and magnetic fields to confine them.

The final phase in these experiments consisted of trapping the antiprotons $\overline{p}$ and the positrons e^+, at a temperature of 4 K (-269 °C), and bringing them into contact so that the so-called *cold atoms* of $\overline{H}$ could form. These atoms move "slowly", which makes it possible for them to be analyzed before

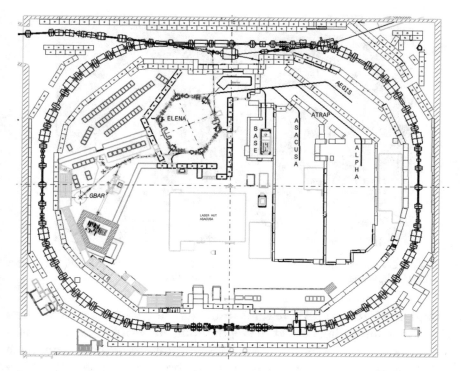

Fig. 8.3 Layout of the hall of the Antiproton Decelerator AD, at CERN, with the different experiments that operate inside the AD ring: ALPHA, ASACUSA, BASE, ATRAP, AEGIS and the new experiment GBAR. The hexagonal structure is the new antiproton decelerator ELENA, coupled to the AD, which is currently connected to all the experiments, expected to resume operation in August 2021. *Credit* Courtesy of CERN

they encounter ordinary matter and annihilate. The positrons for these experiments were obtained from the radioactive isotope of sodium ^{22}Na, which is artificially obtained. The ATHENA experiment ended in 2004, giving rise to the ALPHA experiment as its successor, which would also inherit most of the scientists working on the previous project.

The ATRAP experiment, whose name comes from "Antihydrogen Trap" kept on working, improving its devices and procedures on several occasions. This was the first experiment that used "cold" positrons (instead of electrons) to cool antiprotons; the two were confined in a trap and as a result some percentage of them combined to form atoms of antihydrogen. The ATRAP experiment was also a pioneer in measuring the intensity of the electric field required to ionize the $\overline{\text{H}}$ atoms, that is, to remove the positron from the shell. The results showed that most of the antiatoms so obtained were formed in highly excited states, with the positrons occupying levels from 43 to 55.

Fig. 8.4 View of the interior of the enormous AD hall, at CERN, around 2013, with the ALPHA experiment in front. *Credit* Courtesy of CERN

Until now, the ATRAP research team has not found any differences between the antihydrogen and hydrogen atoms in what refers to the energies of the orbitals. On the other hand, by capturing antiprotons in a Penning trap, the ATRAP experiment has been able to perform high-precision measurements of the charge, mass, and *magnetic moment*[2] of the antiproton, simply by applying static and oscillating voltages to the electrodes of the trap. Its most relevant result, obtained in 2013, consisted in measuring with great accuracy the magnetic moment of the antiproton to compare it with that of the proton, finding that they are identical except for the sign, as they were expected to be. In other words, the ATRAP collaboration found that the magnetic moments of the antiproton and proton are exactly opposite: equal in strength but opposite in direction with respect to the particle's spin.

8.2.2 ALPHA

Now let us see the ALPHA experiment, successor of the ATHENA experiment, whose name comes from "Antihydrogen Laser Physics Apparatus" (Fig. 8.4). It began operating in April 2008 and its main mission is to create, capture and analyze antihydrogen $\overline{H}$ atoms to compare them with hydrogen

---

[2]The magnetic moment of a particle, or an atom, is a property that indicates its sensitivity to magnetic fields.

H atoms. The analysis of antiatoms is done using laser and microwave spectroscopy, as well as other techniques. In June 2011, the ALPHA collaboration was able to trap around 300 $\overline{H}$ atoms for 16 min, enough time to start studying their properties in detail. This result, of capital importance, was due to the fact that the ALPHA apparatus used a magnetic trap to catch antiatoms. Indeed, thanks to the fact that electrically neutral atoms are equipped with a small magnetic moment, they can be confined to magnetic traps, while electric fields only allow to capture ions or charged particles, but not neutral atoms because these lack electric charge.

ATHENA lacked such a magnetic trap and was mixing the antiprotons $\overline{p}$ and the positrons e^+ in a trap only for charged particles. Consequently, once the $\overline{H}$ atoms were formed, being electrically neutral they drifted away annihilating themselves against the walls of the trap. Those walls recorded the annihilations, which could then be analyzed, and it was found that these antiatoms only survived a few microseconds after they were created.

Although neutral atoms can be confined in appropriate magnetic traps, it turns out that these traps are very weak and can only capture $\overline{H}$ atoms with maximum kinetic energies—those associated with their speed—equivalent to a temperature of 1 K (-272 °C), only one degree above absolute zero. To achieve such low temperatures, the ALPHA experiment uses a technique known as evaporative cooling. However, in order to apply the laser spectroscopy techniques, it is necessary that the antiatoms are in their lowest energy state—the ground or fundamental state—with the positron in the 1S orbital, which has not been the case for most antiatoms created until now. This circumstance greatly reduces the number of antiatoms that can be analyzed using these techniques.

At the end of December 2016 the ALPHA collaboration marked a milestone, since it published the first observation of the 1S-2S transition in $\overline{H}$ atoms, which thereby became the first spectral line observed in these antiatoms and the first observed spectral line in antimatter atoms ever. The method followed consisted in irradiating the antiatoms, confined in the magnetic trap, with two opposite beams of ultraviolet laser light of wavelength $\lambda = 243$ nm. This light is resonant with the 1S-2S transition in hydrogen atoms H, because the absorption of two of these head on photons makes the electron jump from the orbital 1S to the orbital 2S (because they travel in opposite directions, the total angular momentum of the photons is zero, and therefore the angular momentum in the transition is conserved). In this way, if the $\overline{H}$ atoms have exactly the same energy levels as the H atoms, exposure to this light will cause a large part of the antiatoms located in the ground state to become excited, their positron jumping from the 1S to the 2S orbital by absorbing a pair of head on photons.

What happens next is that the excited $\overline{H}$ atoms leave the magnetic trap because it does not exert enough strength to retain them as they are too energetic to be held. As a consequence, these antiatoms crash against the trap walls and annihilate themselves (see Fig. 8.5). With this is mind, the researchers irradiated the $\overline{H}$ atoms with both resonant and non-resonant lasers, and then analyzed, in both cases, how many antiatoms had been annihilated against the walls of the magnetic trap and how many remained inside the trap. They found that 67 antiatoms remained in the trap when the laser was adjusted to the resonant λ, compared to 159 when the laser had no such adjustment. And as for the atoms that escaped from the trap and were annihilated against the walls, the researchers were able to detect that 79 of them had done so while being irradiated by the resonant laser, as opposed to 27 when being irradiated by the non-resonant laser. These comparisons made them conclude that the resonant laser light was, in fact, interacting with the antihydrogen $\overline{H}$ atoms via the 1S-2S transition, hence identical to that of the hydrogen atoms.

The ALPHA collaboration improved those results slightly in 2017, and shortly afterwards, in 2018, the team published the first observation in $\overline{H}$ atoms of the Lyman-α spectral line corresponding to the 1S-2P transition in hydrogen atoms (see Fig. A1 in Appendix A), whose wavelength is $\lambda = 121.57$ nm. To accomplish that transition, that is, to jump to the 2P orbital, it is only necessary for the positron of the 1S orbital to absorb one photon with that value of λ, so all it takes is a single laser beam adjusted to that

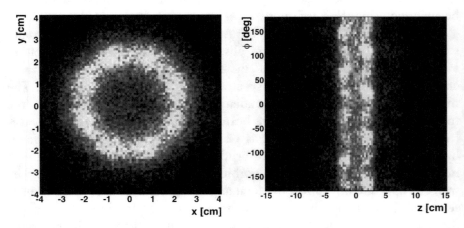

Fig. 8.5 Untrapped antihydrogen atoms annihilating on the inner surface of the ALPHA trap. These are measured by the ALPHA annihilation detector. The events are concentrated at the electrode radius of about 22.3 mm. *Credit* Courtesy of CERN

resonant λ. Curiously, although this requires a simpler experimental setup than the previous experiments, it turns out that in practice, in the laboratory, it is much more difficult to adjust a laser light beam to the wavelength of $\lambda = 121.57$ nm than to deal with the two lasers of the previous arrangement.

Furthermore, in 2017 the ALPHA collaboration had already succeeded in inducing hyperfine atomic transitions within the fundamental level 1S by exposing the $\overline{H}$ atoms to a given microwave radiation. These transitions correspond to energy sublevels very close to each other and originate from the interaction between the spin of the positron and the spin of the antiproton in the atomic nucleus. The team was able to measure two hyperfine spectral lines of the antihydrogen $\overline{H}$ atom and they found no difference comparing them with the homologous hyperfine spectral lines of the hydrogen H atom (within the experimental limits). Other hyperfine transitions followed, within the level 2P, and with the results of all these observations the team was able to deduce for the $\overline{H}$ atoms the homologous of the Lamb effect, which is the energy difference between the levels 2S and 2P with total angular momentum $j = 1/2$. These levels are separated by a tiny energy, 4.372×10^{-6} eV, and the effect is due to the interaction of the electrons, or positrons, with the quantum fluctuations of the electromagnetic field in the vacuum, as predicted by Quantum Electrodynamics. This result was recently published in Nature, in February 2020.

In addition to the techniques using spectroscopy, in 2013 the ALPHA collaboration published some results describing the first direct analysis of how antihydrogen is affected by Earth's gravity, as we will see in detail when we describe the AEGIS experiment. Surprisingly, the researchers attempt to elucidate whether the antiatoms fall into the terrestrial gravitational field with the same acceleration as matter; that is, whether antimatter participates in the universality of free fall. Moreover, there are also some who propose that antimatter might feel gravitational repulsion towards matter instead, so that the $\overline{H}$ atoms would rise rather than fall in the terrestrial gravitational field. Anyway, although the ALPHA team was not even able to determine if the $\overline{H}$ atoms fall down or rise up, and the results they found were of no relevance, this analysis marked a path to follow to achieve better results in the future.

8.2.3 ASACUSA

The ASACUSA experiment (Fig. 8.6), "Atomic Spectroscopy And Collisions Using Slow Antiprotons", also makes extensive use of spectroscopy, as its name suggests, although in a different way from that of the other experiments. It was started at the same time as the ATRAP experiment, in February

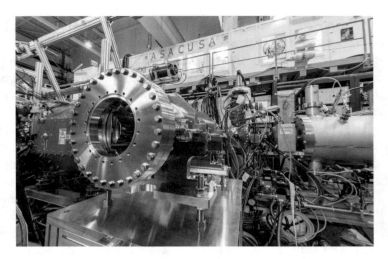

Fig. 8.6 Close-up of the ASACUSA experiment, which investigates antimatter using spectroscopy techniques on antihydrogen atoms as well as on antiprotonic helium atomcules. *Credit* Courtesy of CERN

2002. The ASACUSA collaboration has a strong Japanese presence, five institutions from that country versus five institutions from as many European countries. The very name of this experiment becomes the name of a Japanese Zen temple just by changing the letter C to K. As a matter of fact, during a first scientific meeting before the experiments began, in the Workshop "Atomic Collisions and Spectroscopy with Slow Antiprotons", which took place in Tsurumi in July 1999, the participants were offered a course on Zen Buddhism and they were taken to practice meditation in the temple at 3:30 a.m.

One unique feature of the ASACUSA experiment, which makes it especially curious, is that it uses *antiprotonic helium* for its investigations with laser spectroscopy, besides antihydrogen. The corresponding "atoms" actually have an intermediate structure between atoms and molecules, and consist of helium atoms with one of the two electrons replaced by an antiproton $\overline{p}$, which having a negative electric charge, fits perfectly in place of the electron e^-. We see then that antiprotonic helium is a hybrid in two senses: on the one hand, it is constituted by matter and antimatter, and on the other, the structure of its exotic atoms is halfway between an atom and a molecule because the antiproton $\overline{p}$ is the nucleus of the $\overline{H}$ atom. Therefore, these exotic atoms are composed of two atomic nuclei, like a molecule, but their electronic structure is like that of an atom, which is why they are also called *atomcules* (Fig. 8.7).

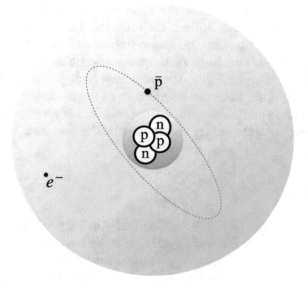

Fig. 8.7 Sketch of an "atomcule" of antiprotonic helium. It is a hybrid between matter and antimatter particles and is halfway between an atom and a molecule since it is composed of two atomic nuclei, like a molecule, but their electronic structure is like that of an atom. Due to its bigger mass, the orbit of the antiproton around the helium nucleus is deep inside the electronic cloud occupied by the electron, and corresponds to a very elliptical orbital

These exotic atoms are produced by mixing antiprotons with helium gas, although most of the antiprotons are immediately annihilated when they come into contact with the protons in the atomic nuclei of the gas. But about 3% of them succeed in ripping an electron from the shell of an atom and orbiting around the nucleus instead, replacing the electron, although in a very elliptical orbital. In this manner, the antiprotons can remain in the atomic shells for tens of microseconds, that is, of the order of 10^{-5} s—which is enough time for this kind of experiments—before finally falling into the nucleus and getting annihilated by a proton.

By using laser spectroscopy on the antiprotonic helium, adjusting the wavelength of the light until the antiprotons resonated and jumped from their orbital to a higher one, the ASACUSA collaboration was able to determine the mass of the antiproton relative to that of the electron, finding that it is identical to the mass of the proton with an accuracy of one part in 10^9. This result was published in July 2011 and was improved slightly in November 2016, by successfully cooling down two thousand million antiprotonic helium atomcules to a temperature of 1.5 K.

In addition, in January 2014 the team achieved to produce a beam of $\overline{H}$ atoms for the first time. Indeed, the researchers detected 80 antiatoms at a distance of 2.7 m from the magnetic trap where they had been created, at a point where the influence of the magnetic fields was negligible. This result opened up the possibility of conducting high precision studies of the antiatoms in flight, far from the interference of magnetic fields. For example, the study of the hyperfine structure using microwave spectroscopy, but with a different and complementary technology to the one used by the ALPHA collaboration.

8.2.4   BASE

The BASE experiment (Fig. 8.8), "Baryon Antibaryon Symmetry Experiment", was initiated in September 2014. It is the only experiment in the AD hall that does not create or work with antiatoms and its aim is to compare, with a high accuracy, the fundamental properties of protons and antiprotons, such as the magnetic moment and the charge-to-mass ratio. To perform the experiments, the researchers cool the antiprotons up to 1 K, and then catch them using sophisticated electromagnetic containers—an advanced Penning trap system—thanks to which they have managed to store an appreciable amount of antiprotons for more than a year. From these containers the antiprotons are supplied one by one to other additional magnetic traps to observe their behavior, which allows to determine their *magnetic moment*.

Fig. 8.8 The BASE experiment investigates properties of protons and antiprotons, like the magnetic moment and the charge-to-mass ratio. *Credit* Courtesy of CERN

This property is one of the most studied intrinsic characteristics of particles, and in theory should be identical for particles and their antiparticles, although with opposite signs, as pointed out before. Failure to be so, would indicate violation of the symmetry *CPT*, which is considered "untouchable".

The most relevant results obtained by the BASE collaboration, published in 2017, have been the most precise determination to date of the magnetic moment of both the proton and the antiproton. The measurements made show that the magnitudes of their magnetic moments are identical, with a fractional precision of 1.5 parts per billion, improving the results obtained by the ATRAP collaboration, in 2013, by a factor of more than 3000. In addition, they have performed the most precise test of CPT invariance, in the baryon sector, by comparing the proton-to-antiproton charge-to-mass ratio with a fractional precision of 69 parts per trillion.

8.2.5 AEGIS

The AEGIS experiment (Fig. 8.9), "Antihydrogen Experiment: Gravity, Interferometry, Spectroscopy", completed installation in January 2013. It is a totally European project, in which 16 institutions from nine countries participate. Its main objective, for the first few years, is to measure the acceleration acquired by antihydrogen $\overline{H}$ atoms when they are subjected to the Earth's gravity alone, and to compare it to the corresponding acceleration acquired

Fig. 8.9 Antimatter trap of the AEGIS experiment, whose aim is to determine the gravitational behavior of antiatoms. *Credit* Courtesy of CERN

by the hydrogen H atoms; what is the same, to study the free fall of anti-matter bodies in the terrestrial gravitational field and to compare it with the free fall of matter bodies. Let us take a closer look at this.

In Chap. 2 we saw that mass is a multifaceted property of particles and bodies in general, and we mentioned three facets of it. Two of these are involved in the free fall of a body in the gravitational field of another one, in which no other force intervenes. These two facets of the mass are known as inertial mass and gravitational mass, which we can write as m_i and m_g, respectively. The first is responsible for the resistance of the body to be accelerated (or decelerated); that is, to change its speed, both in magnitude and in direction or sense, and is expressed by the formula $F = m_i a$. The gravitational mass m_g, on the other hand, acts as the gravitational charge of the body.

Let us remember that, according to Newton's Law of Universal Gravitation, the force of gravitational attraction between two bodies is proportional to the product of their masses and inversely proportional to the square of the distance that separates them. Therefore, since the weight P that makes a body fall into the terrestrial gravitational field is precisely the force of attraction between that body and the Earth, we can express it using Newton's Law in the form:

$$P = G\, m_g \frac{M_T}{r^2}, \tag{8.1}$$

where G is Newton's constant, m_g is the gravitational mass of the body, M_T is the mass of the Earth, and r is the distance between the body and the center of the Earth. In this context we can write the above formula, $F = m_i a$, like:

$$P = m_i\, g, \tag{8.2}$$

where g is the acceleration acquired by the body of inertial mass m_i subjected to the attraction of the Earth; or, in other words, by falling into the Earth's gravitational field. Now, by equating the two expressions (8.1) and (8.2) for the weight P one gets:

$$m_i\, g = G\, m_g \frac{M_T}{r^2}. \tag{8.3}$$

In this equality we see that if the inertial mass and the gravitational mass are identical, that is, if $m_i = m_g$, then the acceleration g produced by the terrestrial gravity is universal, that is to say, it is the same for all the bodies, since it does not depend on their masses but only on the height at which they fall, resulting in:

$$g = G \frac{M_T}{r^2}.$$ (8.4)

The universality of free fall, i.e. the independence of g from the masses of the falling bodies, has been proven with great precision in the case of matter, implying that the inertial mass m_i and the gravitational mass m_g are identical, at least within the known experimental limits. As a matter of fact, if we consider another planet or a star or satellite, instead of the Earth, the results are similar, obviously, simply by replacing in expression (8.4) the mass of the Earth M_T with the mass of that particular celestial body. Interestingly, the astronauts of the Apollo 15 mission recorded a small video on the Moon, available on the internet, where the commander, David Scott, is seen standing on the lunar surface dropping a hammer and a feather simultaneously from the same height. As expected, the two objects reach the ground at the same time due to the absence of air in the Moon. For it is the resistance of the air that impairs this type of experiments here on Earth, forcing us to perform them inside devices from which the air has been extracted.

Returning to antimatter, the experiments designed to study its behavior, under the action of Earth's gravity, have to elucidate three questions directly related to the universality of free fall. One question is if it also extends to antimatter; that is, if the acceleration of gravity g has the same value for both matter and antimatter bodies. If not, the reason could be that the gravitational constant G is different for the attraction between two bodies of matter and the attraction between a body of matter and a body of antimatter. Another question, which is actually a particular case of the previous one, is whether the force of gravity between one body of matter and another of antimatter is also attractive, as implied by General Relativity. If not, this would represent a before and an after in theoretical physics, and would lead to an authentic revolution.

And the last question is whether the inertial mass m_i and the gravitational mass m_g of antimatter bodies are also identical, as in the case of bodies of matter. In fact, it could happen, and this is related to the second question, that these masses are identical in magnitude, but with opposite sign (in a similar way to electric charges with opposite sign). This would result in a negative gravitational mass for antimatter, which would generate repulsive forces on bodies of matter. And conversely, the bodies of matter would generate repulsive forces on antimatter bodies, so that antihydrogen atoms $\bar{H}$ would fall "upwards", instead of downwards, in the terrestrial gravitational field. In Fig. 8.10 the two possibilities are considered, as indicated by the question marks.

The ALPHA collaboration published in 2013 two results on the behavior of antihydrogen in the terrestrial gravitational field, as mentioned before. One of them was that their gravitational mass is less than 110 times their inertial mass, that is, $m_g < 110 \, m_i$, which is a result of little relevance given that these two masses are expected to be identical. The other result was that antihydrogen is not repelled by the Earth with a gravitational mass larger than 65 times its inertial mass, that is, $m_g \leq 65 \, m_i$, in the event of gravitational repulsion between matter and antimatter.

The AEGIS collaboration, on the other hand, has not published any results yet on the gravitational behavior of antihydrogen, although it has published most recently their results on the production of antihydrogen in a pulsed mode, which is the first phase of their experiment . As we said, its mission consists of measuring the acceleration of gravity g with which the $\overline{\text{H}}$ atoms fall down, or rise upwards, under the influence of the terrestrial gravity alone, that is, in free fall far from the reach of any another force. To produce antiatoms, they use radioactive sodium ^{22}Na as source of

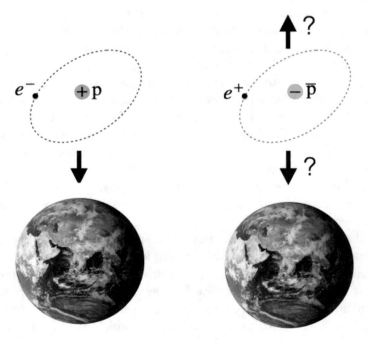

Fig. 8.10 Free fall of atoms of hydrogen H (left) and antihydrogen $\overline{\text{H}}$ (right) in the Earth's gravitational field. Do antiatoms fall down or rise up?

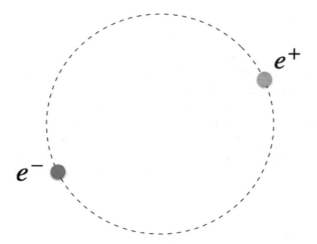

Fig. 8.11 "Atom" of positronium, a bound state of one electron and one positron orbiting around their common center of mass. If their spins are parallel (aligned) it is named orthopositronium and its mean lifetime is 142 nanoseconds. If their spins are antiparallel (oppositely aligned) it is named parapositronium and its mean lifetime is only 125 picoseconds

positrons, and they combine these with the antiprotons that come from the AD decelerator, as is done in the previous experiments. However, the method followed is very different since the researchers first manipulate the positrons using nanotechnology so that they form bound states with the electrons, the so-called *positronium* (Fig. 8.11). This enables the positrons to last longer before they are annihilated by the electrons.[3] Then, they excite the positronium with laser pulses, in order to distance the positrons from the electrons. This procedure has two effects: it prevents these particles from annihilating one another and it loosens the electrical force between them thereby facilitating the capture of positrons by antiprotons. Finally, the researchers fire the positronium pulses against the antiprotons, in such a way that a pulsed beam of $\overline{H}$ atoms is obtained traveling horizontally.

These beams pass through an instrument called a *Moiré deflectometer*, which consists of a system of grids that split them up and transforms them into parallel beams. Finally, those antiatoms that manage to cross all the grids impact against a vertical screen, annihilating themselves with the matter. The exact points where the annihilations occurred, as well as the velocities carried by the antiatoms, can be analyzed with great precision through photographic

---

[3] For reasons that are not well understood, positronium interacts with matter very much like electrons moving at the same speed, as if the positrons were concealed. This effect protects positrons from being annihilated by the surrounding electrons.

emulsions and other devices. And from these data it is possible to deduce how much the $\overline{H}$ atoms have fallen during their flight, which enables to determine the acceleration g of gravity for them.

The experiment is in the phase of analyzing the annihilations of the $\overline{H}$ atoms against the screen, producing, above all, pions and heavy ions. As this task is very arduous and laborious, the AEGIS collaboration has launched the project named *Antimatter* in a *crowdcrafting* platform—tasks carried out by a multitude of people—of citizen science. In particular, the researchers ask all interested people to participate in the analysis of the annihilations by recognizing the different trace patterns left by the pions on the photographic emulsions.

8.2.6 ELENA

The experiments just described have the good fortune that a new instrument has been coupled to the AD decelerator to further reduce the speed of the antiprotons decelerated by this one. This is the ELENA decelerator (Fig. 8.12), with a 31 m ring, whose name means "Extra Low Energy Antiproton". As pointed out before, the AD decelerator can only slow down

Fig. 8.12 The new decelerator ELENA, with hexagonal shape and 31 m long, is coupled to the AD decelerator and will reduce the energy (5.3 MeV) of the antiprotons coming from the AD decelerator by a factor 50, until 0.1 MeV, before delivering them to the different experiments in the AD hall. *Credit* Courtesy of Bert Schellekens

the antiprotons coming from the PS accelerator up to an energy of 5.3 MeV, which corresponds to too high a speed still to allow the antiprotons to be analyzed and to capture positrons to form antihydrogen $\overline{H}$ atoms. For this reason, each experiment has to look for other procedures to reduce the speed of the antiprotons much more, so that their kinetic energy translates into extremely low temperatures, between 1 and 4 K.

This logically affects the efficiency and performance of the experiments, which is why it was decided to build a second antiproton decelerator to be coupled to the AD decelerator. ELENA's ring will reduce the energy of the antiprotons coming from it by a factor of about 50, up to 0.1 MeV. In addition, it will increase the density of the beams, which will improve the efficiency with which the experiments capture the antiprotons in their electromagnetic traps by a factor between 10 and 100. And, on top of this, ELENA will also allow all the experiments to receive antiproton beams simultaneously, what the AD decelerator is unable to do.

In the photo shown in Fig. 8.13 the new and old chairpersons of the AD Users Community (ADUC) are posing at the entrance of the AD hall facility, called *Antimatter Factory*, where one can see the logos of the two decelerators:

Fig. 8.13 The new and old chairpersons of the AD Users Community (ADUC) posing at the entrance of the AD hall facility, CERN's Antimatter Factory, in January 2018. From left to right, in front are Chloé Malbrunot and Stefan Ulmer, recently elected new ADUC chairs, and behind are Walter Oelert and Horst Breuker, the previous ADUC chairs. At the Antimatter Factory entrance one can see the logos of the two antiproton decelerators: curved AD on the left and hexagonal ELENA on the right (designed by Breuker's daughter). *Credit* Courtesy of CERN

a curved AD on the left, and ELENA, showing its hexagonal shape, on the right.

On August 2, 2017, the first antiproton beams of 5.3 MeV circulated in the ELENA ring coming from the AD decelerator. This was a test experiment for its assembly, as important parts for its operation were still missing, especially in the system of radio frequencies used to decelerate the antiprotons. Finally, in April 2018, with all its elements assembled, testing and adjustments to the machine began for commissioning. The hope was that ELENA could go into operation, connected to the new GBAR experiment, in the summer of 2018, before the technical shutdown of all CERN accelerators in middle November. Unfortunately, a major malfunctioning of the AD decelerator, which manifested itself at the beginning of the summer, delayed ELENA's programming until October, resulting in a short debut of the GBAR experiment, which we will see next.

8.2.7 GBAR

The GBAR experiment (Fig. 8.14) was installed in the AD hall in December 2016, and was coupled to the ELENA decelerator a few months later. Its name comes from Gravitational Behavior of Antihydrogen at Rest. It has a

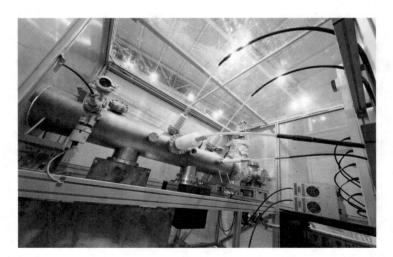

Fig. 8.14 Image of the GBAR experiment. It received the first antiprotons from ELENA in October 2018, but only until November 13, due to the technical shutdown at CERN. It will start operating properly in August 2021, searching for the acceleration of antihydrogen atoms in free fall on Earth's gravitational field. *Credit* Courtesy of CERN

significant French presence—five institutions from that country—along with several European, Japanese, and Korean institutions.

The main objective of this project, like that of the AEGIS collaboration, is to measure the acceleration of gravity, g, experienced by antihydrogen atoms in free fall on Earth, but the techniques used are very different. First of all, the GBAR experiment will not use neutral $\overline{H}$ antiatoms until the last instant, but ions $\overline{H}^+$ with positive electric charge, consisting of two positrons e^+ orbiting an antiproton $\overline{p}$. These ions are very difficult to produce, but once created they can be manipulated much more easily than the neutral antiatoms using electromagnetic fields. Secondly, GBAR needs positron beams with an intensity 100 times greater than that provided by radioactive isotopes. Due to this, they had to build a small linear accelerator 1.2 m long, named GBAR linac, which accelerates electrons to an energy of 10 MeV and makes them collide with a target, producing the required positrons.

In this experiment, positrons will also be manipulated to form positronium with electrons, and then the researchers will excite the positronium with laser light before mixing it with the antiprotons coming from the ELENA decelerator. In the tests that have already been performed, most antiprotons capture a single positron, but some of them capture two positrons (in two consecutive phases), thereby creating the ions $\overline{H}^+$. These will be slowed down much more, until their speeds are less than 1 m/s, equivalent to temperatures of 10 μK (millionths of Kelvin!), which is why they are referred to as ultra-cold ions. Finally, using a laser of appropriate wavelength, it will be possible to remove the additional positrons from the $\overline{H}^+$ ions, and thus convert these into neutral $\overline{H}$ antiatoms, subject only to the strength of the Earth's gravitational field, so that they will fall (or rise?) in it. The first phase of the GBAR experiment will consist in the creation of the $\overline{H}^+$ ions, and the second phase will focus on analyzing the free fall of the neutral $\overline{H}$ antiatoms.

8.2.8 FAIR

We cannot end this chapter without mentioning the FAIR complex, an international laboratory for cutting-edge research with antiprotons and ions, whose name is International Facility for Antiproton and Ion Research. It is currently being built in Hesse, near Darmstadt (Germany) and will become one of the largest particle accelerator facilities in the world. Indeed, these installations will accommodate about 3000 scientists from some 50 countries, and will allow several projects to be carried out at the same time, like at CERN, but using much lower energies for their research. In fact, the main accelerator ring is only 1.1 km long, through which atomic nuclei and ions will be accelerated.

The first relevant experiments at FAIR are expected to be performed by the end of 2025. These experiments are distributed into four main projects, or collaborations, with the names APPA (Atomic, Plasma Physics and Applications), CBM (Compressed Baryonic Matter), NUSTAR (Nuclei and Stars) and $\overline{P}$ANDA (Antiproton Annihilation at Darmstadt).

As their names suggest, these experiments will cover many scientific areas with the aim of exploring the nature of matter and its many facets. For example, the CBM experiment will replicate the conditions inside supermassive objects, like neutron stars, subject to extreme temperatures, pressures, and densities. On the other hand, the research at APPA will range from the study of atoms and materials up to engineering and cancer treatment. As for antimatter, the $\overline{P}$ANDA experiment will use beams of antiprotons to make them collide against protons. The analysis of the products so obtained will give the researchers a better understanding of various aspects of the strong force and the hadrons; in particular, the generation of the hadron masses through the energy produced by that force.

9

Medical and Technological Applications of Antimatter

This last chapter is dedicated to the practical applications of antimatter in our society, which consists basically in the applications of positrons e^+ and, to a lesser extent, of antimuons μ^+, also called positive muons. These applications can be differentiated into two large groups. On the one hand, there is the daily use of positrons in clinics and hospitals to apply the imaging technique known as *Positron Emission Tomography*, which shows the condition of organs and tissues, and serve to locate and study tumors, among many other functions. On the other hand, there is the use of positrons and antimuons at an industrial level, and as a tool in many investigations in the area of Materials Science and Technology.

In what follows, we will describe in some detail some of these applications, which constitute what we could consider *the useful side of antimatter*. First of all, let us remember that positrons are emitted by some radioactive isotopes. Among them, Sodium-22 (^{22}Na) is broadly used as a positron source, for example to produce antihydrogen atoms at the CERN AD hall facility, also called Antimatter Factory, as was pointed out in Chap. 8. In addition, it is easy to create positrons by colliding a beam of electrons or γ rays from a reactor, or a very intense laser beam, against a target. In these cases, electron–positron pairs, e^+e^-, are produced and the particles can be separated before they annihilate each other. The muons, in turn, are created by making a beam of protons hit a target, which produces an abundance of pions that immediately decay producing muons, as was shown in Fig. 2.5 and Eq. (2.3), in Chap. 2. It should also be noted that most of the particle accelerators that exist today are quite small and are used for medical or technological purposes, either in hospitals or laboratories.

© Springer Nature Switzerland AG 2021
B. Gato-Rivera, *Antimatter*,
https://doi.org/10.1007/978-3-030-67791-6_9

To finish, we will explain the reasons why matter–antimatter annihilation cannot be used as a source of energy to provide for the basic needs of our daily life, neither for the propulsion of spaceships nor for constructing bombs (despite the popularity of these ideas in science fiction literature and movies).

9.1   Medical Applications

The annihilation of positrons with electrons from biological tissues constitutes the basis of Positron Emission Tomography (PET). This imaging technique is widely used in nuclear medicine nowadays, as it serves a multitude of purposes. To apply it, some substances called radiotracers or radiopharmaceuticals are injected into the patient. These are chemical compounds in which one or more atoms have been replaced by a short-lived, positron-emitting, radioisotope of elements that are abundant in the body, like Carbon-11 (^{11}C), Nitrogen-13 (^{13}N), Oxygen-15 (^{15}O) and Fluor-18 (^{18}F), the latter being of paramount importance for the localization and monitoring of tumors, as we will see. Since these radioactive isotopes are short lived (some of them last only a couple of minutes before disintegration), they must be produced just before being injected into the patient bloodstream. To do this, the corresponding elements are bombarded with protons coming from a small accelerator.

The patient is then placed inside the PET scanner (Fig. 9.1), or tomograph, whose internal walls consist of a series of detector rings, that record the gamma radiation emitted when the positrons are annihilated inside the body. This radiation consists of two γ rays of 511 keV, for each e^+e^- annihilation, leaving the body in opposite directions (see Fig. 2.2 in Chap. 2)—180 degrees from each other—which allows the detectors to locate the exact point in the body where the annihilation took place. The device scans along the body, or just an area of interest, and the recorded signals are used to make a series of slices that combine to form a 3-D image.

PET scans may detect the early onset of many diseases, simply by identifying tiny changes in the cells, before other imaging techniques can do it. In general, they allow doctors to assess the condition of organs and tissues as they can monitor blood flow and many bodily and metabolic processes, including neuronal transmission. This is also the reason PET scanners are widely used in the pharmaceutical industry to analyze the behavior of medicines in the body.

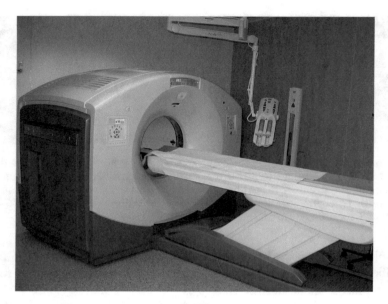

Fig. 9.1 PET scanner from the "PET-Zentrum am Diakonie–Klinikum Stuttgart", Germany. It is actually a PET Computed Tomography scanner, or PET/CT–System. CT imaging uses X-rays and provides excellent images of the inside of the body. The two combined scans provide more accurate diagnoses than performed separately. *Credit* Courtesy of user Hg6996 and Wikimedia Commons

Moreover, PET scans are especially suitable to study the structure and activity of the brain, being able to even track the physiological changes associated with diseases like alcoholism and depression. In fact, this technique has played a major role in the study of neurotransmitters. It was the first imaging technique suitable for the brain, starting in the mid-1970s, and provides images with spectacular color contrasts, where the warmer colors represent the more active areas of the brain (Fig. 9.2). This is because the brain's cognitive functions affect the blood flow. Indeed, when a group of neurons becomes more active, the capillaries surrounding them dilate automatically in order to deliver more blood (more oxygen) to them. As a result, during the PET scans the dilation of the capillaries in the more active areas of the brain gives rise to the emission of more positrons, and therefore of more gamma radiation.

Furthermore, PET scans are also heavily used to obtain fast and precise 3-D images of tumors, allowing to follow tumor growth and to search for metastases, therefore helping doctors to plan the oncological therapies that should be administered to the patients. The radioisotope ^{18}F is of crucial importance because it allows locating the tracer *Fluorodeoxyglucose* (18FDG)

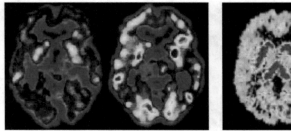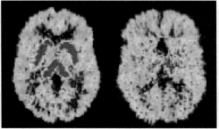

Fig. 9.2 On the left, one can see PET scans of the brain of an alcoholic patient 10 days (left) and 30 days (right) after starting the abstinence cure. On the right, PET scans of the brain of a healthy person (left) and a person with severe depression (right), revealing low levels of the neurotransmitter serotonin. *Credit* McGill University, Canada

after injecting it into the body, thanks to which glucose consumption can be identified and quantified, making it possible to detect tumor cells due to their elevated glucose metabolism.

A curious fact is that in organic molecules positrons combine with electrons, before their mutual annihilation, forming positronium "atoms", which we already encountered in Chap. 8 (see Fig. 8.11). This happens more than 80% of the time when irradiating biological tissues with positrons and increases the effectiveness of the PET techniques by retaining the positrons for a few moments, thus increasing the likelihood of their annihilation in the desired area. Positronium is a bound system with energy levels similar to those of the hydrogen atom, although the wavelengths of the corresponding spectral lines are roughly twice the wavelengths of the analogous hydrogen lines.

The lowest energy level of positronium—its ground state—comes in two possible forms, or configurations, depending on the relative orientations of the spins of the electron and the positron. One is *ortho-positronium*, with the electron and positron spins parallel, which decays producing three photons, having a mean lifetime of 1.42×10^{-7} s. The other configuration is *para-positronium*, with the spins aligned in opposite directions, which decays preferentially into two γ rays with an energy of 511 keV each, having a mean lifetime of 1.25×10^{-10} s. This is precisely the e^+e^- annihilation detected by the PET scanner.

As for the use of antimatter therapy to destroy tumors, the convenience of using antiprotons $\overline{p}$, instead of protons, has been under investigation for several years (protons are used in some radiotherapy treatments, usually combined with chemotherapy). One of the advantages of irradiating tumors with antiprotons would be the drastic reduction of the necessary doses of radiation, since the antiprotons would be annihilated inside the cancer cells releasing much more energy than that achieved with proton bombardment.

The first investigations carried out to evaluate the effectiveness and suitability of antiproton radiation in oncological therapies took place at CERN. This was the objective of the ACE experiment (Antiproton Cell Experiment), in which cells were irradiated with antiprotons. This experiment was installed in the AD hall of the Antiproton Decelerator, from 2003 until its completion in 2013, bringing together a team of experts in Physics, Biology and Medicine from 10 scientific institutions. Hence these researchers became the pioneers in the study of the biological effects from irradiation with antiprotons. In their experiments they compared cell damage using antiprotons versus protons and concluded that only a quarter of antiprotons were needed to cause the same level of cell destruction.

9.2   Technological Applications

The annihilation of the electrons of matter with positrons provides a most useful tool in Materials Science and Technology, as we said. Positive muons μ^+ are also used in some techniques, although on a smaller scale. These do not annihilate with electrons, as is obvious, but they decay spontaneously producing positrons that do annihilate with the electrons of the surrounding matter. Now, let us have a brief look at some of the technological applications of antimatter.

PET tomography is employed to obtain images of the interior of some materials, especially for the purpose of analyzing processes of technological or industrial interest. For example, the flow of water or oil through rocks, the behavior of detergents in washing machines and dishwashers, the behavior of lubricants in engines, the state of radioactive waste in nuclear deposits, etc. It can also reveal the onset of metallic fatigue much sooner than any other technique, which is why it is used to examine the state of turbine blades in airplanes, allowing to improve flight safety.

There is also *Positron Annihilation Spectroscopy* (*PAS*), which consists of irradiating samples with low energy positron beams and making highly accurate measurements of the γ ray pairs emitted when the positrons are annihilated with the electrons in the samples, often after positronium is produced. This technique is most suitable to analyze in detail the physical and chemical properties of the surfaces and thin films of many materials, allowing the study of condensed matter microstructure at a scale as small as a few tenths of nanometers (see Fig. 9.3). PAS spectroscopy comprises several types

Fig. 9.3 Device designed to magnetically confine positron beams for PAS (*Positron Annihilation Spectroscopy*) experiments. It is used by the Slow Positron Research group in the University of Bath for the characterization of the physical, chemical, and electronic properties of solids on the nanometer scale. *Credit* Slow Positron Research group, University of Bath (UK)

of measurements. One of them is based on the Doppler broadening of the energy spectrum of γ rays due to the surrounding atomic environment. It can follow changes with temperature, for example to study ice films, and is suitable for detecting structural defects, like vacancies of atoms and pores in the crystalline structure of many solids. The variation in angles between the emitted γ rays, on the other hand, provides very precise information about the electronic and magnetic properties of materials of technological interest such as high-temperature superconductors.

Another technique of PAS spectroscopy is that associated with lifetime measurements of positrons or positronium atoms when they are implanted into materials; that is, the time they last before being annihilated with the surrounding electrons. This time is related to the electron density of the material at the annihilation site, and, in the case of positronium formation in the material, the ortho-positronium lifetime is related to the free volume size. This procedure gives information about the size and properties of open defects in materials. It is used to study complex oxides with technologically important properties, such as ferroelectricity, superconductivity, and magnetoresistance, as well as study semiconductor materials for solar panels and other electronic devices. In addition, lifetime measurements are also used, for the sake of the pharmaceutical and food industries, to probe open cavities of sub-nanometric scales (smaller than 10^{-9} m) in biopolymers, with the objective of designing tailor-made vehicles for the transport of medicines and nutrients inside the blood stream.

Positive muons μ^+, or antimuons, are also used as probes to investigate the properties of many materials. Interestingly, by irradiating those materials with antimuons, a kind of hydrogen-like atom is formed, called muonium (Fig. 9.4), with an antimuon in place of the proton, and with an electron stripped off from the material. But these exotic atoms disintegrate with a mean lifetime of 2.2×10^{-6} s when the antimuons decay giving positrons as by-products:

$$\mu^+ \rightarrow e^+ + \nu_e + \bar{\nu}_\mu, \tag{9.1}$$

which are annihilated with the electrons in the material, thereby allowing their detection. The electron neutrinos ν_e and muon antineutrinos $\bar{\nu}_\mu$ which are also produced in the μ^+ decays, as shown in (9.1), pass completely unnoticed.

Alternatively, the muon spin spectroscopy called μ S R, *muon Spin Rotation*, is a technique based on the resonance of the magnetic moment of the

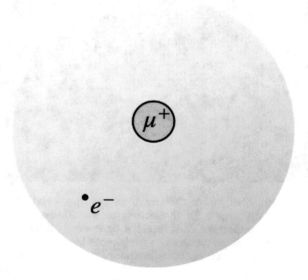

Fig. 9.4 The muonium atom, with an antimuon in the nucleus and an electron in the shell, disintegrates when the antimuon decays, according to (9.1), with a mean lifetime of 2.2×10^{-6} s. Its energy levels are similar to those of the hydrogen atom

antimuon, analogous to the well-known *Nuclear Magnetic Resonance*. It is most suitable for the study of magnetic fields on an atomic scale inside materials, being especially useful for the analysis of the structure of compounds with novel or potentially valuable electronic properties. This technique is also appropriate for studying the role of hydrogen impurities in semiconductors, superconductors, and exotic molecular magnets.

To finish this section, it is worth mentioning that large scale production of positrons has been performed in several labs for different purposes. For example, in 2008 a research team from the Lawrence Livermore National Laboratory in California (USA), led by Hui Chen (Fig. 9.5), produced more than 100 thousand million positrons in a laser experiment using short pulses. For doing this, the researchers shoot their ultra intense "Titan" laser onto a thin - one millimeter thick - gold sample. Using a very powerful laser as well, in 2013 another team, from Michigan University, succeeded in building a device about one meter long capable of generating short bursts of both electrons and positrons in enormous amounts (10^{15}). This effect is similar to what is expected to be emitted by the jets from massive black holes and pulsars (see Sect. 4.6.2 in Chap. 4).

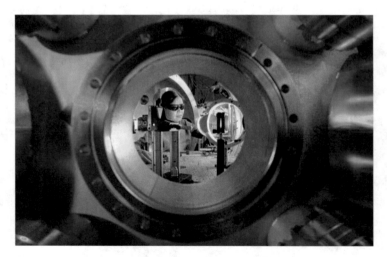

Fig. 9.5 Physicist Hui Chen, from the Lawrence Livermore National Laboratory in California (USA), sets up targets for the antimatter experiment at the Jupiter laser facility. She has made important contributions to Plasma Physics, most notably in the field of relativistic positron generation via intense laser-matter interactions, and she is the leader of these studies, both in theory and experiment. She designed, built, and calibrated the electron, positron, and proton spectrometers needed for this research and developed the data analysis methods. *Credit* Courtesy of Hui Chen and Lawrence Livermore National Laboratory

9.3   Antimatter as an Energy Resource?

The application of matter–antimatter annihilation to cover the energy needs of our society—in our cities, factories, etc.—similar to the use of nuclear energy, can still take more than a century of technological evolution, if this can ever be achieved at all. The same can be said of its relevance for the propulsion of rockets and spacecraft, and of its possible uses in the war and military industry, fortunately.

The fundamental reason for this is that with our current technology is only possible to produce negligible amounts of antimatter, compared with what would be required to meet our most basic energy needs. And furthermore, those amounts of antimatter, however small (for example, to illuminate a room), would be obtained at the cost of a gigantic effort, both energetic and financial, because to use antimatter as a source of energy one first has to create it. Needless to say, if some natural antimatter reservoirs were found and we had access to them, all the energy problems of our society would be solved. Let us now look at some figures to put this question into perspective, which will allow us to appreciate, in passing, the uncontrolled imagination of some fiction writers.

First of all, one should bear in mind that in order to produce a single antiproton $\bar{p}$ with the aid of accelerators, we have to accelerate about five million protons until they reach very high energies, moving at near the speed limit c, and then make them hit a metal target. In addition to this, one finds that if we were able to store all the antiprotons that were produced at CERN for one year, of the order of 10^{13}, their annihilation with the same number of protons would supply only the energy needed to keep a light bulb of 100 W lit for 30 s. This can be easily verified simply by taking into account that the energy released in a $p\,\bar{p}$ annihilation is, approximately, 3×10^{-10} J and one *watt* can be expressed as 1 W = 1 J/s, where J is the symbol of the *joule*, the unit of energy in the *International System of Units (SI)*.

It must be noticed, however, that the proton-antiproton annihilation, $p\,\bar{p}$, is not "clean" in the sense that these particles do not simply disappear giving rise to photons, as is the case for the electron–positron annihilation, e^+e^-, at low energies, as shown in Fig. 2.2 of Chap. 2. Rather, their annihilation purely into photons is suppressed by a factor of about 3×10^{-7} and the dominant processes consist of annihilations into pions plus neutrinos[1] and photons. The pions decay very quickly producing more photons and neutrinos, as well as e^+e^- and $\mu^+\mu^-$ pairs, the latter decaying in turn

---

[1] Here and in what follows the term neutrino refers to both neutrinos and antineutrinos.

producing electrons, positrons and more neutrinos, as shown in (9.1) for the case of the antimuon. If this were not enough complication, when an antiproton is put into contact with matter and interacts with an atomic nucleus, the antiquarks of the antiproton may annihilate with the quarks of the protons, but also, and equally well, with the quarks of the neutrons. In other words, *antiprotons annihilate not only with protons but also with neutrons*, and in both cases the by-products include many particles besides photons, especially neutrinos, except in the special case of the $p\,\overline{p}$ annihilations which are purely electromagnetic, i.e. purely into photons.

Anyway, if we now examine the data from the largest antiproton factory that has ever existed, whose main purpose was to supply the antiprotons to the Tevatron at Fermilab (Chap. 6, Fig. 6.10), we see that in its most prolific years it succeeded at producing some 10^{15} antiprotons per year, sufficient to bring to a boil a liter of water at room temperature if they were completely annihilated with the same number of protons. To be precise, 1.4×10^{15} $p\,\overline{p}$ annihilations would be required to raise the temperature of one liter of water from 0 up to 100 °C (one has to take into account that the heat needed to raise the temperature of one gram of water by one degree °C is one calorie and equals 4.18 J).

Secondly, let us remember that one gram of antimatter when annihilated with the same amount of matter would produce an energy equivalent to more than double the energy released in the explosion of the Hiroshima atomic bomb, as noted in Chap. 2. However, since one gram of antiprotons is equivalent to 6.02×10^{23} of them, it would take 602 million years to produce a single gram of antiprotons using the technology of the Fermilab factory (which is no longer available because it was decommissioned at the closure of the Tevatron in September 2011). As for CERN, using its current technology it would require 60,200 million years to produce one gram of antiprotons, more than four times the 13,800 million years that the Universe has existed. If this production rate could be increased by a factor of one hundred, which is not at all evident, it would reach the rate of the Fermilab factory. This rate of antiproton production is also the expectation for the $\overline{P}$ANDA experiment, at the FAIR facility.

But the problems just mentioned are not the only ones, since the storage of large quantities of particles that repel each other, having the same electric charge, whether antiprotons or positrons, is also a major undertaking and fraught with difficulties. Let alone that antimatter particles would annihilate against the walls of their container as soon as any contact occurred between them. To minimize these difficulties, the alternative procedure would be to store antiatoms, which are electrically neutral. But it would still take hundreds of millions of years to create one single gram of antihydrogen atoms at the expected rate of production at the CERN Antimatter Factory (Fig. 8.13), which is the only facility in the world where antiatoms are created nowadays.

Matter–antimatter Annihilation versus Nuclear Bombs

Let us consider, again, the example of the bomb that exploded over Hiroshima on 6 August 1945, dubbed Little Boy, whose destructive power was 15 kilotons of TNT, approximately. Taking into account that:

$$1 \text{ kiloton of TNT} = 10^{12} \text{ calories, and } 1 \text{ calorie} = 4.18 \text{ J}, \qquad (9.2)$$

one finds that the energy released by Little Boy was 63×10^{12} J. Dividing this quantity by the energy produced in one $p\,\bar{p}$ annihilation, which is 3×10^{-10} J, one finally obtains that 2.1×10^{23} $p\,\bar{p}$ annihilations are necessary to match the energy of that bomb:

$$63 \times 10^{12} \text{ J} / 3 \times 10^{-10} \text{ J} = 2.1 \times 10^{23}. \qquad (9.3)$$

But one gram of antiprotons corresponds to 6.02×10^{23} of them, a number almost three times larger. In conclusion, the annihilation of one gram of antiprotons with one gram of protons would release almost three times the energy produced by the Hiroshima bomb.

Appendix A: Atomic Spectroscopy

This appendix explains some basic notions about atomic spectroscopy, in particular applied to the hydrogen atom H. This technique is an essential tool used in several experiments carried out at CERN, as described in Chap. 8, in order to analyze the properties of antihydrogen atoms $\overline{H}$ as well as those of the atomcules of antiprotonic helium. Atomic spectroscopy is also an essential tool in Astronomy, Astrophysics and Cosmology since it allows us to identify the different chemical elements and molecules that take part in the composition of stars, galaxies and interstellar clouds of gas, together with their velocities with respect to the Earth. This is due to the fact that atoms of different elements, as well as different molecules, all have different absorption and emission spectral lines that faithfully characterize them, like fingerprints, and these spectral lines get blue or red shifted according to the velocities of these objects with respect to us, as explained in Chap. 3.

Atomic spectroscopy allows us to investigate the internal structure of atoms and it consists of bombarding them with photons of different wavelengths λ; that is, photons with different energies, since these are inversely proportional to λ. In this way, it is possible to observe the excitations of the electrons (or the positrons, in the case of antiatoms), since they jump to higher energy levels by absorbing photons of appropriate energies, descending shortly afterwards with the emission of photons as well, either of the same or of related wavelengths. Let us have a closer look at this.

Unlike classical physical systems, like planets revolving around a star, atoms are quantum systems. This implies, among other things, that their energies do not take continuous values but only discrete specific ones. Thus, while planets can change their trajectory and orbit in a continuous manner by taking any

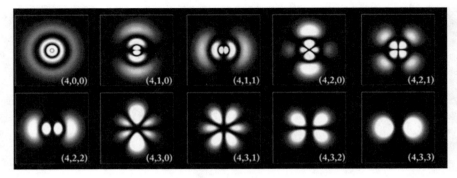

Fig. A1 Illustration of the atomic orbitals of the electron in the hydrogen atom at level 4. The notation (n,l,m) indicates the quantum numbers for the energy level, the angular momentum, and the magnetic moment. Orbitals of type S and P have l = 0 and l = 1, respectively. Quantum mechanics cannot predict the exact location of the electron, only the probability of finding it when detected. Here, the brighter areas represent a higher probability of finding the electron. *Credit* Courtesy of PoorLeno and Wikimedia Commons

values, as a result of various disturbances, electrons of the atomic shells cannot move around the nuclei in whatever way, and are distributed in space forming the so-called atomic *orbitals*, with permitted energy levels. These orbitals are very different from the classic orbits, as they actually are configurations resembling electronic "clouds" that indicate the probability of finding the electrons when they are detected (see Fig. A1).

To rise to a higher level, electrons need a precise energy to make that quantum leap, called *transition*. Therefore, if they are irradiated with *resonant photons*, that is, of the energy they need, they will be able to make the transition by absorbing one or more of those photons. Then the electrons will find themselves in an *excited state*, although not for long because electrons tend to return to their previous state by emitting the excess energy also in the form of photons, as if they returned the ones that they had absorbed. As a consequence, all the elements of the Periodic Table, of which hydrogen is the simplest, and also all molecules,[1] have a characteristic spectrum of absorption and emission of photons, as mentioned above.

The orbital with the lowest energy level is the orbital 1S, where 1 indicates the level and S denotes that it has spherical symmetry. But this orbital only admits two electrons, with opposite spins, which means that the other electrons must be placed at higher levels. Since the hydrogen atom has only one electron, the atom is at its lowest energy level, called the ground state,

---

[1] Molecular spectra are much more intricate, however, due to the internal motion inside molecules, and consists of three types of spectra: electronic, rotational and vibrational.

when the electron is located in the orbital 1S. If excited by resonant photons, then the electron can jump to the 2S, 3S, or even higher orbitals, and also to non-spherical orbitals of the same levels. But they descend to a lower level shortly thereafter giving off one or more photons, whose energies sum up exactly the difference of energy of the electron between the two levels of the transition. Furthermore, in very energetic environments, like stars, many of the electrons which are excited at higher levels get only partially de-excited, descending to the orbitals of level 2 or 3, or even higher, before returning to the orbital 1S or becoming excited again by absorbing more photons.

The transitions of the electron of the hydrogen atom from the upper levels to the orbital 1S, emitting a single photon, produce the spectral lines known as *the Lyman series*, for it was Theodore Lyman who studied them between 1906 and 1914. All these lines, numbered as α, β, γ, etc., are found in the ultraviolet region of the electromagnetic spectrum (see Fig. A2). The orbitals from which these transitions are made are not spherical, though, but the so-called orbitals 2P, 3P, 4P, etc. This is so because the emission, or absorption, of a single photon by an electron cannot produce a transition between two spherical orbitals. The reason is that in atomic transitions the angular momentum is preserved, as well as the so-called parity. The former is equal to zero in spherical orbitals, while the photon spin (its intrinsic angular momentum) is equal to 1 and, in addition, the photon reverses parity. Hence, the Lyman-α line corresponds to the transition 2P-1S from the orbital 2P to the orbital 1S; the Lyman-β line corresponds to the transition 3P-1S; the Lyman-γ line corresponds to the transition 4P-1S, etc.

The transitions that end up in the orbital 2S, on the other hand, have been well known since the nineteenth century because four of these lines are within the visible spectrum of the sunlight. They are referred to as *the Balmer series* in honor of Johann Jakob Balmer, who in 1885 found the mathematical formula that accurately described these visible spectral lines. Obviously, neither Balmer nor his contemporaries could understand the reason for that formula, and it still took many years before Quantum Physics came up with the answer. As it turned out, this answer unfolded in two phases: first in 1913 through the atomic model of Niels Bohr (fairly rudimentary), and almost two decades later through Quantum Mechanics. The transitions ending in the orbitals 3S and 4S are named the *Paschen and Brackett series*, respectively, and their lines are located in the infrared and far infrared range of the electromagnetic spectrum. In Fig. A2 one can see a scheme of the first spectral series of the hydrogen atom.

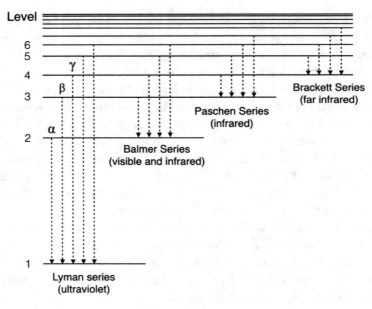

Fig. A2 The first spectral series of the hydrogen atom

Appendix B: The Myth of Skobeltzyn and the Positron Tracks

The aim of this appendix is to debunk a myth about the discovery of the positron, involving the Russian physicist Dmitri Skobeltzyn[2] (Fig. 5.2), that seems to be spread mainly among British physicists. First we will go through the facts, most of which were already reviewed in Chap. 4, and especially in Chap. 5, and then we will expose the fiction, which began to take shape in the middle 1950s in the mind of a philosopher of science named Norwood Russell Hanson. To finish, we will point out two errors in the book of Gordon Fraser "Antimatter", from the year 2000, that are also spreading and on their way to become legends. So, we hope to arrive on time to stop this from happening.

The myth in question consists of the false belief that Skobeltzyn had seen, but not recognized, positron tracks much earlier than 1930, and therefore several years before Carl Anderson's discovery of the positron, that happened in 1931 although it was only confirmed one year later, in 1932, after Anderson came up with the "lead plate test". About the myth, Skobeltzyn himself declared more than once that he saw positron tracks for the first time only in early 1931, probably prior to any other scientist, but he ignored how to interpret these tracks until Anderson published the positron discovery one year later.

One word of caution: when we used the expression "to see positron tracks" in the previous paragraph it was implicit that, technically, these tracks could

---

[2]Also spelled as Skobeltsyn and Skobelzyn.

© Springer Nature Switzerland AG 2021
B. Gato-Rivera, *Antimatter*,
https://doi.org/10.1007/978-3-030-67791-6

be distinguished from electron tracks. For it they could not, then the discussion that followed does not make any sense. As a matter of fact, the cloud chamber inventor Charles Wilson surely was the first person to see tracks of cosmic-ray particles—and therefore of positrons and muons—in the 1920s or even before, but it was impossible for him to identify them for the simple reason that his cloud chambers were not placed in a magnetic field. This means that all the traces observed and recorded by Wilson were straight lines, from which it is impossible to deduce the electric charge and the energy of the particles.

Most of the facts that follow are documented in contributions by Skobeltzyn himself to three books and one article. One is the chapter *The Early Stage of Cosmic Ray Particle Research* in the book "The Birth of Particle Physics", from 1983. This book is based on the lectures and round-table discussions that took place at the "International Symposium on the History of Particle Physics", held at Fermilab in May 1980. In this congress, Skobeltzyn presented his reminiscences on the early stage of cosmic ray research, and he recounted the history of the discoveries from the middle 1920s to the middle 1930s. This chapter was almost exactly reproduced in the book "Early History of Cosmic Rays Studies", from 1985. But there is an important difference between the two versions of this chapter: in the first book, Skobeltzyn added a crucial note at the end of the chapter which is of great relevance, as we will see.

The other contributions of Skobeltzyn to the facts are excerpts and comments from his correspondence with Norwood Russell Hanson, mainly from two letters that Skobeltzyn wrote in 1956 and 1960. Some of those were presented by Hanson in the 1961 article *Discovering the Positron (I)*, published in "The British Journal for the Philosophy of Science". This article was published again in 1963 in the book "The Concept of the Positron", by the same author, together with some additional relevant information about the correspondence between Hanson and Skobeltzyn, in Appendix IV.

The Facts

In the autumn of 1923, Dmitri Skobeltzyn, working in the laboratory of his father in the Leningrad Polytechnical Institute (USSR), built a Wilson cloud chamber to study the Compton effect, which had just been discovered. This effect consists of the scattering of energetic photons, such as X-rays or γ rays, with the electrons of matter (see the footnote 1 of Chap. 5). He obtained γ rays from a radioactive source of Bismuth-214, dubbed RaC. During the

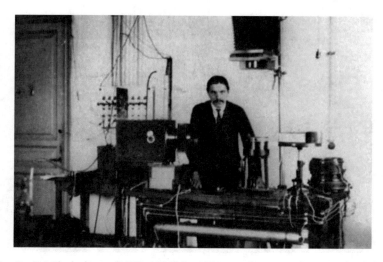

Fig. B1 Dmitri Skobeltzyn (1892–1990) in 1924, at the age of 32, in the laboratory of his father in Leningrad, with his Wilson cloud chamber (near his left arm) placed in a magnetic field of about 1000 gauss. He was the first scientist to put a cloud chamber in a magnetic field and in this apparatus he made the first observations ever of the Compton effect of γ rays using such a device. *Credit* Soviet Academy of Sciences

first months he was taking photographs of the electron tracks he wanted to study, produced in the gas of the chamber (the γ rays, lacking electric charge do not leave any traces, but they can be studied following their impact on the electrons with which they interact). However, there was a problem with the visibility of the images, produced by large amounts of unwanted electrons hanging around, which were created by the impact of the γ rays against the walls of the Wilson chamber.

Then, a few months later, in 1924 a brilliant idea occurred to him. This was to place the cloud chamber in a magnetic field of about 1000 gauss to deflect the unwanted background electrons (Fig. B1). In this improvised way one of the most important scientific apparatus was born: *the Wilson cloud chamber operated in a magnetic field.* Indeed, this instrument proved to be crucial for the further developments in Nuclear Physics and for the birth of Particle Physics as an independent discipline.

Shortly afterwards, Skobeltzyn decided to use a stronger magnetic field, ranging from 1500 to 2000 gauss (Fig. B2), and during 1925–26 he took more than 600 stereoscopic images. Among them, 27 showed straight tracks that could only be interpreted as belonging to highly energetic "electrons", traveling at near the maximum speed c and unrelated to the γ rays from the radioactive sources. That is, these "electrons" had to be produced by

Fig. B2 In 1925–1926, using this setup with a stronger magnetic field (1500–2000 gauss), Skobeltzyn recorded straight tracks of very energetic particles, that he rightly ascribed to cosmic rays. Analyzing these straight tracks, he was the first scientist to show that cosmic rays can be highly energetic particles. *Credit* Soviet Academy of Sciences

the cosmic radiation entering the laboratory, and they were so fast and energetic that the magnetic field was unable to bend their trajectories. As a result, it was not possible to distinguish the sign of the electric charge of the particles, but this was not a problem for Skobeltzyn as he, like all scientists at that time, believed that the charged particles in the cosmic radiation were all electrons, usually referred to as β rays. Only ten years later, four types of particles were known that could have produced the straight tracks of Skobeltzyn: electrons e^- , positrons e^+ , muons μ^- and antimuons μ^+ .

In 1927, Skobeltzyn published, for the first time, photographs with straight tracks in an article in "Zeitschrift für Physik" (Fig. B3), and two years later he published another article with more details in the same journal. But in the meantime, he presented the pertinent facts and photographs during the "International Conference on γ and β Ray Problems", in Cambridge (UK), from July 23 to 27, 1928. On that occasion, Skobeltzyn showed a collection of photographs of the cosmic-ray straight tracks that produced a strong impression on the audience; so strong that these images were the spark that ignited the cosmic ray research using cloud chambers operated in powerful magnetic fields capable of bending the cosmic-ray trajectories.

In early 1931, working on the Compton effect with a radioactive source of ThC', Skobeltzyn saw traces of positrons for the first time; to be precise, of electron–positron pairs. However, he did not understand the

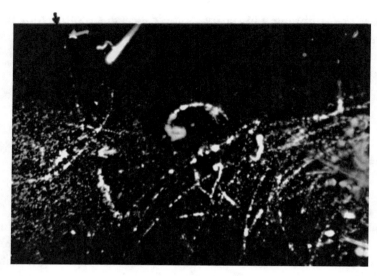

Fig. B3 One of a stereoscopic pair of photos, published by Skobeltzyn in "Zeitschrift für Physik" in 1927, showing a straight track that was produced by a highly energetic particle, indicated by one black and two white arrows. This was the first cosmic-ray straight track discovered by Skobeltzyn in 1925–1926 using the arrangement in Fig. B2. The many curved tracks belong either to the "Compton electrons" of the gas that interacted with the γ rays from the radioactive source of ^{214}Bi (RaC), or to spurious background electrons kick out from the walls of the chamber by the γ rays. *Credit* "Zeitschrift für Physik" and Soviet Academy of Sciences

meaning of the bizarre traces similar to those of electrons but bending the other way around—showing positive electric charge—until one year later, when Anderson published the discovery of the positron. Now, according to Skobeltzyn, this was the first time ever that positron tracks were seen (although not identified), since no other scientists had seen positron tracks before him, i.e. positron tracks that could be distinguished from those of electrons.

So far so good. But now comes another episode that, although partly based on true facts, seems to have been incorrectly reproduced about 20 years later by Paul Dirac, presumably due to inaccurate reminiscences. In October 1934, Skobeltzyn attended the "International Conference on Physics" in London, where he participated in the *Discussion on Cosmic Radiation*. Most meetings of the conference took place in London, spread in different rooms of different institutions, but a group also met at Cambridge on Thursday, October 4, by invitation of Lord Rutherford (this is so indicated in the preface of the Conference Proceedings). That day, after the Conference sessions, Skobeltzyn was invited to a private meeting held in the evening in a room of one of the colleagues at Cambridge University. Some colleagues wanted him to speak about his recent observations of positrons (he had presented a short communication on the subject at a plenary session in London), and Dirac was in the audience.

The Fiction

Norwood Russell Hanson

Norwood Russell Hanson (1924–1967) was a philosopher of science and a Professor of History and Logic of Science in Indiana University (USA). Although he was a U.S. citizen, from New York, he went with his wife to the U.K. in 1949, with a Fulbright Scholarship, where he completed several degrees at the universities of Oxford and Cambridge. Eight years later, Hanson left Cambridge to return to the U.S., in 1957, founding the Indiana University Dept. of History and Philosophy of Science. But before that, in the middle 1950s, he decided to investigate the genesis of the discovery of the positron and to write a couple of articles and a book about it.

Now, and this is very important, it happened that Hanson already had a pre-conceived hypothesis in mind, related to his previous work, that he wanted to prove. Namely, that important discoveries technically attainable could however be overlooked for cultural or psychological reasons. In short, he wanted to prove that the positron had been "seen but not observed" (using Hanson's terminology), several years earlier than Anderson's discovery, in 1932, because the researchers were not mentally prepared for it. As a matter of fact, there was some truth in this hypothesis, since something of this kind actually occurred to several scientists in 1931–32, including Skobeltzyn, Millikan, and the Joliot-Curies, but not before that (see Chap. 4 for more details). The problem was that Hanson badly wanted to demonstrate that the positron had been "seen but not observed", earlier than in 1930.

With this idea in mind, in 1955 Hanson started gathering information and contacting scientists, mainly experimental physicists, both in person and through correspondence. His strategy was to find out whether any of them had seen positron tracks in cloud chambers operated in a magnetic field. The latter was a necessary ingredient because without a magnetic field the traces of positrons would be identical, undistinguishable from the traces of electrons. So, Hanson started searching for positron tracks, that in fact had to look very similar to electron tracks but curling 'the other way around' by the action of the magnetic field. This behavior produced visual effects in the cloud chamber photographs, described in different ways, like electrons: 'moving backwards' or 'falling back into the radioactive source', or 'curving the wrong way' or 'coming upwards from the floor' (in the case of cosmic rays).

Unfortunately, at the very beginning, it seems that Hanson was led astray by Dirac, who pointed to Skobeltzyn as a candidate to have seen positron tracks much before 1930. This is what Hanson wrote about a conversation with Dirac around 1955, in the article *Discovering the Positron (I)*, and in his book "The Concept of the Positron":

> Prof. P. A. M. Dirac once spoke to me of a lecture given at the Cavendish by D. Skobeltzyn, 'sometime in 1926 or 1927'. Dirac recalls the description by someone, perhaps Skobeltzyn, perhaps someone in the audience, of an experimental setup within which Skobeltzyn was bombarding a metal target. One of the curiosities reportedly mentioned by Skobeltzyn was that several particles which were certainly electrons were seen to 'fall back into the source'; this, despite the fact that most of the electrons moved in the way usual for this experiment, *away* from the source. Professor Dirac feels that what he remembers Skobeltzyn as having then described could only have been positive electrons, and he suggests that the Russian might very well then have made the discovery.

This paragraph, provided Dirac's words were correctly reproduced by Hanson, which is rather dubious (see below), would show that Dirac confused, by seven or eight crucial years, the date of his attendance to the Skobeltzyn evening talk in 1934, where the speaker was actually discussing his observations of positrons in γ ray experiments. However, Skobeltzyn would not have used in his talk any expression like " *...electrons were seen to 'fall back into the source'....*", as he himself pointed out to Hanson, suggesting that Dirac could have confused his talk with another one of F. Joliot-Curie in 1932 (page 138 in Hanson's book). Moreover, the last sentence about Dirac suggesting that the Russians might have discovered the positron is very hard to believe, really unthinkable. Also to be noted is that Dirac was difficult to approach in person because he talked very little. Thus, it seems rather strange that he was so talkative with Hanson, as the latter suggests in his paragraph.

My sincere impression is that most of that paragraph, written four or five years after the conversation with Dirac, could have been Hanson's own imagination, based on a much shorter conversation with him. One also has to take into account that, at the time Hanson wrote the paragraph, he was totally convinced that Skobeltzyn had seen tracks of positrons around 1926 (or at least he behaved as if he was), and this conviction could certainly have influenced his memories. In addition, Hanson was not very reliable, as the following example shows.

In his book, in page 141 one finds Fig. 4 (p. 200 of the article, Fig. 3). That figure shows two sketches of two cloud chamber photographs with the relevant tracks. According to Hanson, these two photographs were published on 18 December 1931, under the title: *Cosmic Rays Disrupt Atomic Hearts*. However, we have discussed at length the publication with such a title in Chap. 4, Sect. 4.2, and it turns out that Hanson makes an incorrect statement about its content and about the date of publication, which was 19 December. That publication consists of one cloud chamber photograph— not two—together with the portrait of Carl Anderson. We have shown these two images in Chap. 4, in Fig. 4.4 and in Fig. 4.7 (left). Hence, it might well have happened that it was not so much Dirac who messed things up, but Hanson.

Anyway, following that conversation with Dirac, Hanson undertook the endeavor of finding positron tracks in the photographic plates made by Skobeltzyn in 1925–26 and published in 1927–29. These images had been taken two by two in a stereoscopic way; that is, from two different cameras in order to appreciate the important details in three dimensions. For this reason, the images were always presented two by two in the publications. After having examined all the published photographs of Skobeltzyn, Hanson said the following (page 137 in the book and, similarly, in page 196 in the article):

> An exploration of Skobeltzyn's published work entirely failed to reveal any photograph, or even a passing remark, relevant to Dirac's narrative. However, a related photograph did turn up in Zeit. Physik. 43 (1927), 362. Schematically reproduced in Fig. 2 is part of the lower left-hand photograph on that page.

Accordingly, Hanson reproduced in Fig. 2 in his book (Fig. 1 in his article), a sketch of some tracks of the photograph in question, shown in Skobeltzyn's article. In addition, in the book (but not in the article) the piece of the actual photo that contained the tracks was also shown (near page 137). In this photograph (see Fig. B4), one sees two tracks that seem to start (or end) at the same point. However, on its stereoscopic partner (Fig. B5), presented by Skobeltzyn next to the other, the coincidence is not that clear, as can be appreciated with the naked eye. Despite this, Hanson decided to show the "good" photograph in Fig. B4, while hiding the "dubious" partner in Fig. B5, to many experimentalists around (presumably in Oxford and Cambridge), asking their opinion if such tracks could reveal an electron–positron pair. In his own words:

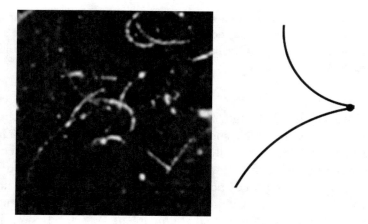

Fig. B4 Detail of the photograph made by Skobeltzyn in his cloud chamber, published in Zeit. für Physik in 1927, lower left-hand photograph on page 362, with the special tracks that so much fascinated Norwood Russell Hanson. According to him, the tracks starting (or ending) at the same point, reproduced in the sketch on the right, corresponded to an electron-positron pair. However, on the stereoscopic partner, shown in Fig. B5, that interpretation is far from obvious. In addition, Skobeltzyn analyzed the stereoscopic composition of the two photographs and concluded that these tracks were unrelated and seemed to overlap just by chance, as shown in the drawing made by Skobeltzyn himself, shown in Fig. B5   *Credit* "Zeitschrift für Physik" and Soviet Academy of Sciences

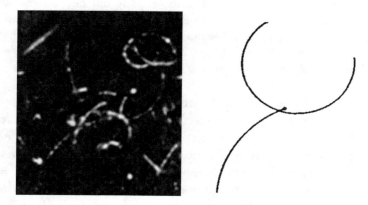

Fig. B5 Detail of the photograph made by Skobeltzyn in his cloud chamber, published in Zeit. für Physik in 1927, lower right-hand photograph on page 362. This is the stereoscopic partner of the image shown in Fig. B4. In this photograph, the interpretation given by Norwood Russell Hanson of the tracks—as an electron–positron pair—is not at all obvious. In addition, the stereoscopic composition of the two photographs revealed that these tracks were unrelated and seemed to overlap just by chance, as shown in the drawing made by Skobeltzyn, on the right. *Credit* "Zeitschrift für Physik" and Soviet Academy of Sciences

As a check on my own reactions this photograph was shown to a great many experimentalists. Without exception, all granted that this might well be a pair; some thought it a very good example of one. No stereoscope was used in this informal survey, however.

Exactly, Hanson concealed the image of the stereoscopic partner, Fig. B5, because it was not good enough, even though both images were published by Skobeltzyn in the same article and in the same page, just next to each other. Observe that, with this procedure, Hanson was inserting in the collective mind of many physicists in Oxford and Cambridge the idea that Skobeltzyn had seen, but not recognized, positron tracks around 1926. And not only that, since Hanson was inquiring about many other physicists at that time in the fields of Nuclear Physics and the incipient Particle Physics, in order to verify that they had seen, but not identified, positron tracks before 1930. Skobeltzyn explained to him that this was not the case, as Hanson relates in his book:

> Skobeltzyn disagrees with my contention that several microphysicists saw, but did not observe, positron tracks prior to 1930. He writes: "None of them published their results earlier than in 1930. Besides, Williams and Terroux used in their work RaE, a source which does not emit γ rays. Meitner-Filipp published their observation in 1933 and certainly did not use the Wilson chamber combined with a magnetic field until 1932. Ellis never worked with a Wilson chamber. Joliot-Curie began the observation with a Wilson chamber placed in a magnetic field only in 1932 (certainly not earlier than the end of 1931, i.e. definitely after my departure from Paris in August 1931).

Skobeltzyn, in addition, informed Hanson in two letters, of 1956 and 1960 (shown in the Appendix IV of Hanson's book), that the stereoscopic composition of the two photographs revealed that the tracks in question were two unrelated tracks that seemed to overlap just by chance. Skobeltzyn even made a drawing himself to illustrate Hanson, shown in Fig. B5. The reaction of Hanson to Skobeltzyn's explanations was: "*Without doubt, Skobeltzyn is correct, but since the original plates are destroyed, the use of a Pulfrich Stereocomparator is not now possible*". And Hanson said this after Skobeltzyn had tried to convince him, with very good manners and enormous patience, that his proposal was completely misleading. For example, in the letter of 10 October 1956, from Moscow, Skobeltzyn wrote:

> The photograph in Zeitschrift für Physik (1927, p. 362), an enlarged reproduction of which is enclosed in your letter, is certainly a result of near coincidence (overlapping) by chance of a track originating in the gas of the cloud chamber

and a spurious track originated without the gamma-rays beam that was under observation. Unfortunately, during the war I lost practically all of my original photographic plates. I have only the print copy of the photo in question in my possession which cannot be examined now stereoscopically with sufficient resolution by means of Pulfrich stereocomparator... And yet it seems clear to me that if one examines carefully this photo even with poor resolution, ... one can see that it is a case of near coincidence by chance and not a real pair. But I repeat that this case as well as many others were carefully investigated during my research work by means of Pulfrich stereocomparators.

And in the letter of 22–24 October[3] 1960, after Hanson had sent Skobeltzyn the proposed redaction of the pages about the pre-discovery of the positron, for his book, Skobeltzyn replied with some rather negative comments. He asserted that many statements in that chapter were erroneous, explaining him with much detail the reasons for it. Moreover, Skobeltzyn also sent to Hanson an enlarged print of the stereoscopic pair of photographs, Figs. B4 and B5, together with the drawing he made, shown in Fig. B5, because Hanson intended to show the "good" photograph (Fig. B4) in his book, while concealing the bad one (Fig. B5), as he finally did. Skobeltzyn wrote:

> ...I am also sending the enclosed reproduction of the photograph (of 1927) that attracted (in vain!) your attention. I am trying once more to convince you that the whole story based on your assertion that the related tracks are tracks of an electron-positron pair is a mistake. I beg you to examine this photo...in a stereoscope......one has to deal with an overlapping by chance of two different tracks - one of a Compton recoil-electron generated in the gas of the cloud chamber and the other a spurious track emerging from the wall of the cloud chamber (if one examines carefully the stereoscopic picture one sees clearly that this is just the case)....

Among several other comments, Skobeltzyn concluded his "tutorial" to Hanson with the very interesting remarks that follow:

> In conclusion I come back again to your point that the positrons could be discovered (and indeed were observed) long time before Dr Anderson's work. I feel (and I must state this bluntly again) that the emphasis on this point in your book leads to a gross distortion in the presentation of the real historical evolution in this field".
>
> "The late C. T. R. Wilson certainly saw the tracks of cosmic-ray particles long before my experiments without being able to identify them (and one can

---

[3]In the headings of the letter one sees the date 22 October 1960, but at Skobeltzyn's signature at the end he wrote 24 October.

find the relevant cases in his beautiful collection of photos published in 1922 or 1923). A simple device (a magnetic field combined with Wilson chamber) used in my experiments for other purposes revealed the nature of these tracks as connected with cosmic rays.

The results of my observations with a relatively weak magnetic field paved the way for the outstanding work of C. Anderson who used a much stronger magnetic field - the main prerequisite of his future success.

It is true that as early as in 1931 (but not earlier as you suggested) I, prior to others, observed electron-positron pairs not being able to identify them, however. (My note in Nature, V. 133, p. 23, 1934.) But no one else contrary to what you state did it at that time or earlier. At the same time C. Anderson, however, had already obtained his first results. It was his excellent experimental technique that in due course led him to his great discovery. But in achieving it he was perhaps somewhat slow. Here your psychological considerations are quite plausible.

As it turned out, Hanson refused to listen to Skobeltzyn and continued to spread his views about the "good" photograph (with the sketch he made), both in the article of 1961 and in the book, published in 1963. Four years later, in April 1967, Hanson died in an accident at the age of 42. One of his passions was flying, and he was killed when his airplane crashed in dense fog near Cortland, New York. Skobeltzyn, by contrast, lived until the age of 98.

But we are not done with Hanson yet, because in the book "The Birth of Particle Physics", published in 1983, Skobeltzyn added a crucial note at the end of his chapter, written around 1981, twenty years after he sent his last letter to Hanson:

"Note added in proof by D. V. Skobeltzyn concerning *The Concept of the Positron* (Cambridge University Press, 1963) by N. R. Hanson:

The interpretation of my results of 1926–27 by the author of this book, based on his examination of plates printed in Zeitschrift für Physik, is nothing else but sheer nonsense. It is absurd to pretend that such conclusions as his can be drawn on the basis of such material. To persuade him that he is wrong, I sent him the original print of a stereoscopic pair of photos that could show him clearly his error. Everybody who has the least habit of exploring such photos would be certainly satisfied. But he persisted in his interpretation. The main point of my remarks is the following: Professor Hanson was inspired by a story told him by Professor Dirac. The story in itself is correct. Dirac remembered a private meeting held in the evening in a room of one of the colleagues at Cambridge University (I remember him attending it) where I had been invited by someone (if I am not mistaken, by the late Cockroft).... A curious point that makes the whole story anecdotal is that Dirac, remembering after a time of about 20 years correctly about the event itself, has been mistaken in indicating its date as something that happened in 1926–27 (see the book by Hanson, p. 136). In fact, it occurred in October of 1934, at the time of the Congress on Physics organized by late Professor R. Millikan.... This shift by six or seven years of the event quoted by Professor Hanson makes his deductions pertaining to it senseless".

Unfortunately, at the time Skobeltzyn wrote that note, around 1981, the absurd story about him seeing positron tracks much before 1931 had taken shape in the collective mind of many British physicists, mainly from Oxford and Cambridge. Some of them had actually met Hanson, who showed them the "good" photograph of Skobeltzyn while hiding from view its stereoscopic dubious partner, and some others did not meet Hanson, but they heard the story....

David Wilson

Curiously, in the same year that the book "The Birth of Particle Physics" was published, 1983, another book was also published where the story of Skobeltzyn and the positron tracks experienced a new twist. This book, "Rutherford, Simple Genius" was written by David Wilson (1927–2000), a science journalist and writer, who for twenty years was the Science correspondent of BBC TV News. To prepare his book, he was able to interview quite a number of scientists who knew and worked with Rutherford in the 1930s, as he explains in the Introduction, page 7. In particular, he mentions F. Dainton, P. Dee, T. Allibone, A. Ratcliffe, A. Kempton, D. Shoenberg and N. Mott.

In this book, in page 562 he is reviewing important achievements by the Cavendish Laboratory in 1932. Then he starts recounting the discovery of the positron by Carl Anderson, and some paragraphs afterwards he explains its immediate confirmation by Blackett and Occhialini at the Cavendish. But in between Anderson and Blackett he slips in the surprising comment:

> Furthermore a number of mysterious cloud chamber photographs had been in circulation among the fraternity of physicists for three or four years: a Russian named Skobelyzyn had shown such tracks at an international conference in Cambridge in 1928, tracks which puzzled everyone including Chadwick, for they seemed to show an electron "going backwards the wrong way".

So, fifty-five years later, the straight cosmic-ray tracks that Skobeltzyn presented in Cambridge in 1928, which initiated the cosmic ray research using powerful magnetic fields, had turned into *curly tracks going backwards the wrong way* (i.e. positron tracks). Hanson had won! In addition, it was a double victory because the only cloud chamber photograph he had found with an apparent positron track (Fig. B4)—that unique photo that he was showing around to physicists in Cambridge and Oxford—had magically replicated and turned into several!

Frank Close

In 2009 the book "Antimatter" was published by Frank Close, a Professor of Physics at Oxford University, in which there is another twist to the plot about Skobeltzyn and the positron tracks. According to the author, already in 1923 traces of positrons were visible in cloud chambers, i.e. traces that could be identified, although nobody became aware of that. Moreover, those traces were produced by both cosmic-ray positrons as well as positrons created in experiments with radioactive substances.

In Chap. 4, Close explains the discovery of the positron, starting in page 49 with the phrase: *"In a nutshell the story is that as early as 1923 the very first photographs of cosmic rays showed images left by positrons, but at the time no one realized the fact".* However, in 1923 nobody had a cloud chamber placed in a magnetic field, and therefore traces of cosmic rays could not be identified, let alone traces of positrons. Skobeltzyn, who was the first to put a cloud chamber in a magnetic field, obtained the very first photographs of cosmic rays that could be identified in 1925–26, not in 1923. But those cosmic-ray tracks were straight, not curved, so it was impossible to distinguish between electrons, positrons and muons. As for the positrons created by γ rays in experiments with radioactive substances, nobody saw any such tracks until early 1931, as Skobeltzyn explained in very much detail. Now let us inspect a few excerpts of the book, from pages 50–52.

> Positrons had been seen, but not recognized, five years before Dirac's theory appeared. In 1923 Dmitry Skobeltzyn in Leningrad was investigating gamma rays; to make them visible he was using a cloud chamber. (p. 50)

The following refers to the moment when Skobeltzyn put his cloud chamber in a magnetic field (actually, in 1924) to improve the visibility:

> This thinned the clouds, and the clearer view revealed something utterly unexpected: the magnetic forces seemed to make some of the 'electrons' curve 'the wrong way'. Today we know that he was seeing positrons, the positively charged 'anti'-version of electrons, but none of this was anticipated in 1923. (p. 51)

Next comes the version of Close about the photographs that Skobeltzyn presented at the Cambridge conference in 1928. It should be noticed that the paragraph shown above, written by David Wilson in his book, is the only reference about Skobeltzyn given in the book by Close, therefore the latter gives essentially the same information:

> News about these images spread among the scientific fraternity and five years later Skobeltzyn decided to show them at an international conference in Cambridge. Everyone was as surprised as he had been, but no one could offer an explanation. It was ironic that he was displaying these images in 1928, in Cambridge, the same year and the same place that Dirac... (p.51)

Yes, everyone was surprised, but because of the straight cosmic-ray tracks presented by Skobeltzyn, denoting that cosmic rays were highly energetic particles, so energetic that the magnetic field used by him was uncapable of deflecting the trajectories of those particles. For this reason, it was technically impossible to tell their electric charge, and therefore to distinguish between electrons and positrons.

Beatriz Gato Rivera

In the period from late 2016 until early 2018, I spent most of my time writing the book "Antimateria" (in Spanish) for the collection of outreach books of the CSIC (Spanish National Research Council), to which my Institute IFF belongs. Among the literature I consulted, I must say that I greatly benefited from reading the books "Antimatter", by the late Gordon Fraser, from year 2000, and the book "Antimatter", by Frank Close, from 2009. As the reader might guess, I found the story presented by Close about Skobeltzyn and the positron tracks so interesting and charming, that I decided to include it in my book. Something was a bit strange, anyway, and perhaps I should have taken it as a warning signal. This was that Gordon Fraser, who had carried out an impressive historical documentation for his book and for the positron discovery in particular, did not mention Skobeltzyn. Furthermore, when I searched on the internet about that story, all the citations, without any exception, referred only to the book of Close as their source.

In 2020, when I started working for the present, very enlarged English version of the previous book, I decided to spend more time than before going through the original literature as much as possible, through both scientific articles and books. Then, when I tried to gather more information about the charming story of Dmitri Skobeltzyn and the positron tracks, I found compelling evidence that this was an invented story that had its roots in past events in Oxford and Cambridge in the middle 1950s. As it turned out, the American science philosopher Norwood Russell Hanson, living eight years in the UK, first in Oxford and later in Cambridge, had "contaminated" the collective mind of many British physicists with unsubstantiated rumors about Skobeltzyn seeing positron tracks much before 1931. I am sorry to have confused the readers of my book "Antimateria" with that false story, and I apologize for that.

Other Antimatter Myths Approaching the Horizon

As mentioned above, Gordon Fraser, who was the editor of CERN COURIER for many years, produced an impressive historical documentation for his book "Antimatter". Unfortunately, two mistakes slipped in, and both of them are now on their course to become myths, slowly but relentlessly. In the first chapter of his book, in page 3, Fraser was wondering if there could be in the Universe worlds where the atomic nuclei had negative electricity and were surrounded by positively charged "electrons". Then he writes:

> In a letter to the journal *Nature* in 1898, the physicist Arthur Schuster surmised 'If there is negative electricity, why not negative gold, as yellow as our own?' For thirty years, Schuster's conjecture gathered dust.

Not really for thirty years (Fraser, believing Schuster meant antimatter, was referring to the Dirac equation of 1928). Schuster conjecture is still gathering dust, after more than 120 years, for he meant *gravitationally repulsive* matter, which is not known to exist.[4] In Chap. 4, Sect, 4.1, we explained the interesting fact that Schuster coined the term "antimatter", and what its meaning was, and we also showed some of the excerpts in his article. Now let us see a few more lines of the Schuster's paragraph from which Fraser extracted the quote shown above:

> If there is negative electricity, why not negative gold, as yellow and valuable as our own, with the same boiling point and identical spectral lines; different only in so far that if brought down to us it would rise up into space with

---

[4]The existence of any kind of matter with repulsive gravitation would contradict General Relativity, so it is not expected to exist. However, as explained in detail in Chap. 8, at the CERN Antimatter Factory some experiments have been designed in order to determine whether antimatter falls in the Earth's gravitational field, like ordinary matter, or rises up.

an acceleration of 98 I. The fact that we are not acquainted with such matter does not prove its non-existence; for if it ever existed on our earth, it would long have been repelled by it and expelled from it. Some day we may detect a mutual repulsion between different star groups, and obtain a sound footing for what at present is only a random flight of the imagination.

The mistake of confusing the Schuster "antimatter" with the Dirac antimatter has been repeated in recent times, and I have spotted it in an article about matter–antimatter domains, written by a cosmologist, and in a very good outreach book written by a particle physicist.

The second error in Fraser's book, in page 71, consists of the assertion that the editor of "Science News Letters", who published the first photograph of a positron track on 19 December 1931, coined the name 'positron'. This is far from true, and reading Fraser's unsubstantiated statements about this episode one gets the impression that he could have got confused talking to somebody at CERN's cafeteria. That error was later reproduced in Frank Close's book, in page 54. However, it should be noted that Fraser mentions Close in the acknowledgements of his book. Therefore, it is not at all clear which of the two confused the other.

As we have seen in Chap. 4, the name 'positron' was suggested by Carl Anderson himself in his 1933 article *The Positive Electron* in "Physical Review", meaning a contraction of 'positive electron'. Anderson stated this in page 943, left, where he wrote: ... *the positive electron which we shall henceforth contract to positron....* Then the editor accepted Anderson's suggestion and stated in the abstract: *These particles will be called positrons.* By contrast, the editor of the "Science News Letters" of 1931 (Watson Davis), only said about the track in question that it was due to a *positive charged particle, probably a proton.* This track is shown in our Fig. 4.7, in the photo on the left, and was marked by Anderson with the letter B.

Epilogue

In the first instants of the existence of the Universe, matter and antimatter particles, which most likely were created in identical quantities, almost totally annihilated each other. Fortunately, just before the Great Annihilation, something happened that gave rise to a very slight excess of matter over antimatter, sufficient for the material Universe to take shape and come into existence as we know it today. Indeed, from the observational data one deduces that for every primordial proton and electron that survived, more than one thousand million succumbed to extinction, along with the same number of primordial antiprotons and positrons. As a consequence, the primordial protons and electrons were immersed in a bath of thousands of millions of photons, for each of them, which 380,000 years later gave rise to the Cosmic Microwave Background radiation, CMB, that we observe since it was discovered in 1964.

Interestingly, the Standard Model of Particle Physics has built-in mechanisms that might produce a tiny excess of matter over antimatter at the beginning of the Universe, but that excess is several orders of magnitude smaller than needed to explain the amount of existing matter.

The vast majority of antimatter particles (or all) that are found in space, which come exclusively in the form of antiprotons and positrons, are not primordial but secondary; at least, their abundances can be explained as such. This means that those antiparticles are mainly produced in collisions of ordinary matter particles with interstellar gas or dust and in very energetic astrophysical processes, especially in the neighboring regions surrounding neutron star pulsars and massive black holes. In addition, there are also proposals that antimatter particles could be created through less understood,

© Springer Nature Switzerland AG 2021
B. Gato-Rivera, *Antimatter*,
https://doi.org/10.1007/978-3-030-67791-6

and not confirmed, phenomena, such as dark matter decay and/or annihilation. Nevertheless, the possibility that a very small amount of primordial antimatter survived in the early Universe cannot be excluded, perhaps one antiproton and positron for every million protons and electrons. In this event, it is not unthinkable that antistars and even small antigalaxies could exist, as some theoretical models predict, provided they were isolated enough from matter so that no detectable signals of annihilation would be produced. In addition, in our observable Universe there are huge, extremely empty regions that might separate large matter domains from small antimatter ones.

The AMS experiment, installed at the International Space Station since May 2011, strives to find clues for such a possibility by scrutinizing the outer space in search of atomic antinuclei (cosmic rays consist essentially of atomic nuclei devoid of electrons). For the moment, after nine and a half years of operation, the AMS collaboration has not confirmed the detection of any antimatter particles besides antiprotons and positrons, but it has collected about ten candidates for antihelium nuclei, and the researchers hope that around 2024 it will be clarified if these candidates are truly antihelium nuclei. Now, according to the experts, if a single nucleus of antihelium were found, this would constitute a major event because it would strongly suggest that primordial antimatter did not totally disappear. This is so because the production of antihelium nuclei by means of very energetic astrophysical processes is expected to be extremely difficult, essentially impossible. Moreover, if the AMS detector were to find bigger antinuclei, such as an anticarbon nucleus (with six antiprotons and several antineutrons), this would be the definitive proof of the existence of antistars since those antinuclei could only have been created in the nuclear furnaces of antistars.

There are also proposals to explain the matter–antimatter asymmetry in the Universe by postulating that particles and antiparticles were not created in identical quantities from the beginning. An interesting possibility, in the theoretical framework of the multiverse, is that our Universe could have been created at the same time as another sibling, almost twin universe, although with the abundances of matter and antimatter exchanged. In this way, the total amounts of matter and antimatter would be identical for the two universes as a whole, but in one of them matter would predominate while in the other antimatter would do so. In this respect, it is worth noting that the scientific possibility of the existence of other universes has already been broadly discussed for several decades both in Particle Physics and in Cosmology.

Now let our imagination run wild and go into the future, to an epoch where we could perform interstellar travel efficiently; for example, through spacetime shortcuts across other dimensions. Suppose we discover an anti-matter star, judging by the antineutrinos it emits (instead of neutrinos), and we resolve to approach it and look closer. Once in the vicinity of the antistar, we see a whole planetary system around it with some of the planets located in the habitable zone. So, we decide to send signals that denote our intelligent origin to those planets, and for this purpose we choose some sequences of laser light flashes with the odd numbers: 1, 3, 5, 7, 9... If we are close enough to those planets, for example two light hours away, it will take only two hours for our signals to reach their destination.

To our amazement, about six hours later we receive an intelligent response from one of those antiplanets, consisting of other sequences of laser light flashes, but this time with the even numbers: 2, 4, 6, 8... This means, without any doubt, that our signals have been intercepted by intelligent beings from a technological civilization made of antimatter, and they send us an acknowledgment using similar signals, although not identical to ours, so that we do not confuse them with an echo of our own signals. Very enthusiastic, next we send to them a friendly video showing the Earth and its people, to which they respond with another friendly video showing us their antiplanet and inviting us to visit them, as can be deduced from their non-verbal language and gestures.

Obviously, they ignore that *we are made of antimatter,* in relation to the matter that constitute them and their entire world. But we do know that they are made of antimatter, in relation to our own matter. Therefore, we cannot accept the invitation and we must restrict the contact exclusively to the interchange of electromagnetic waves. Hence, no official receptions would be possible since in their antiplanet we could not breathe the anti-air, neither we could drink their anti-water nor eat their anti-food; we could only explode, inflicting a great deal of damage to our hosts. And conversely, anti-matter intelligent beings could not breathe our air, neither drink our water nor eat our food. In fact, an advantage of contacting a technological civilization of antimatter rather than one of matter, is that they would never attempt to invade our planet, provided they were aware of our nature, of course. Otherwise, after sending the first spacecraft, they would quickly realize that it is not a good idea to send the rest of the fleet.

In addition, in the unlikely event of a conflict with such a civilization, the armament would be very economical since just throwing a few rocks would be enough to wipe out any enemy antiplanet (remember, the annihilation of only one gram of matter with one gram of antimatter would produce a blast almost three times that of the Hiroshima bomb). But there would be an inconvenience: the armament would be equally affordable to the enemy...

Returning to the present, the writing of this book was completed in November 2020, less than one year before the resumption of the experiments with antiatoms at the CERN Antimatter Factory, after almost three years of being shut down. These experiments will count with the assistance of the new antiproton decelerator, ELENA, which will allow them to be performed much more efficiently than using only the antiproton decelerator AD. A new phase opens, therefore, in the production and analysis of antimatter atoms, in the quest to verify whether they have exactly the same properties as the matter atoms. Moreover, the antihelium candidates observed by the AMS collaboration might give us a big surprise four to five years from now. Stay tuned!

Further Reading[5]

Books on Antimatter

Caballero Carretero JA (2013) *Dirac. La Antimateria*. National Geographic, RBA (Spanish, translated to other languages). Note: The main focus of this book is Dirac and his work

Close F (2009) *Antimatter*. Oxford University Press. Note: I would recommend this book to anybody interested in all the different aspects of antimatter, although it contains some passages that are not correct, as we have explained in Appendix B. Also, the author mentions the possibility of antimatter stars, but does not seem to realize that they would emit antineutrinos, instead of neutrinos

Fraser G (2000) *Antimatter. The ultimate mirror*. Cambridge University Press, Cambridge. Note: This book is very well documented with respect to the historical facts involved in the discovery and use of antimatter particles, especially at CERN. Like the book of Close, this is a must-read for anybody interested in all the different aspects of antimatter. It contains a couple of mistakes that we have clarified in Appendix B

Gato Rivera B (2018) *Antimateria*. CSIC/Los libros de la Catarata (Spanish). Note: This book is the previous, much shorter, version of the present book, as explained

---

[5]Here there is a list of publications related to the present book. First come the popular books that I have read on the subject of antimatter, together with some comments about each of them. Then other popular books are listed, on Particle Physics and Cosmology, which I find especially useful for the layperson. Afterwards, there is a list with most of the articles and (non-popular) books that are mentioned in the different chapters of the book, together with a few more articles that were not mentioned but are also especially interesting.

© Springer Nature Switzerland AG 2021
B. Gato-Rivera, *Antimatter*,
https://doi.org/10.1007/978-3-030-67791-6

in the Preface. It contains a passage that is not correct, as we have mentioned in Appendix B

Hall N (2013) *Antimatter*. Report from the IOP (Institute of Physics). This is a very useful booklet that gives in a nutshell all the important facts about the different aspects of antimatter

Other Books on Particle Physics and Cosmology

Casas A (2009) *El LHC y la frontera de la física*. CSIC/Los libros de la Catarata (Spanish)

Casas A (2010) *El lado oscuro del Universo*. CSIC/Los libros de la Catarata (Spanish)

Fernandez-Vidal S, Miralles F (2013) *Desayuno con partículas*. Plaza & Janes Editores (Spanish, translated to other languages)

Feynman R (1985) *QED: the strange theory of light and matter*. Princeton University Press

Freese K (2014) *The cosmic cocktail. Three parts dark matter*. Princeton University Press

Garfinkle D, Garfinkle R (2008) *Three steps to the Universe: from the Sun to black holes to the mystery of dark matter*. Chicago University Press

Greene B (2011) *The hidden reality. Parallel universes and the deep laws of the Cosmos*. Alfred Knopf Publisher

Hewitt PG (2001) *Conceptual Physics*. Addison-Wesley, New York. Note: Very good introductions to Special and General Relativity

Levin J (2002) *How the Universe got its spots*. Princeton University Press

Masip M (2016) *Los rayos cósmicos*. Colección 'Un paseo por el COSMOS', RBA (Spanish, translated to other languages)

Pais A (1982) *Subtle is the Lord. The science and life of Albert Einstein*. Oxford University Press

Pastor Carpi S (2014) *Los neutrinos*. CSIC/Los libros de la Catarata (Spanish)

Pastor Carpi S (2017) *La Nucleosíntesis*. Colección 'Un paseo por el COSMOS', RBA (Spanish, translated to other languages)

Penrose R (2016) *Fashion, faith and fantasy in the new physics of the Universe*. Princeton University Press

Perlov D, Vilenkin A (2017) *Cosmology for the curious*. Springer, Berlin

Randall L (2005) *Warped passages: unraveling the mysteries of the Universe's hidden dimensions*. Harper Collins: Ecco Press, New York

Rees M (2001) *Our cosmic habitat*. Princeton University Press

Ruiz Lapuente P (2019) *La aceleración del Universo*. CSIC/Los libros de la Catarata (Spanish)

Sobel D (2017) *The glass Universe: the hidden history of the women who took the measure of the stars*. Harper Collins Publishers

Weinberg S (1977 and 1993) *The first three minutes*. Basic Books

Wilczek F (2008) *The lightness of being: mass, ether, and the unification of forces*. Basic Books

Articles Related to Chapter 2

Cheng H, Feng JL, Matchev KT (2002) *Kaluza-Klein dark matter*. Phys Rev Lett 89:211301
Feynman R (1949) *The theory of positrons*. Phys Rev 76(6):749
Randall L, Sundrum R (1999) *Large mass hierarchy from a small extra dimension*. Phys Rev Lett 83(17):3370

Articles Related to Chapter 3

Brax P, van de Bruck C (2003) *Cosmology and Brane Worlds: A Review*. Class Quantum Gravity 20(9):R201
Crispino L (2020) *The October 10, 1912 solar eclipse expeditions and the first attempt to measure the light-bending by the Sun*. Int J Mod Phys D 29:2041001
Einstein A (1916): *Die Grundlage der allgemeinen Relativitätstheorie* (The Foundation of the General Theory of Relativity). Annalen der Physik 49:769
Elizalde E (2019) *Reasons in favor of a Hubble-Lemaître-Slipher's (HLS) law*. Symmetry 11:35
Freese K (2005) *Cardassian expansion: dark energy density from modified Friedmann equations*. New Astron Rev 49:103
Lemaître G (1927) *Un Univers homogène de masse constante et de rayon croissant rendant compte de la vitesse radiale des nébuleuses extra-galactiques* (A homogeneous Universe of constant mass and growing radius accounting for the radial velocity of extragalactic nebulae), Annales de la Société Scientifique de Bruxelles 47:49
Steinhardt PJ (April 2011) *The inflation debate*. Scientific American 304:48

Articles Related to Chapter 4

Anderson CD (1932) *The apparent existence of easily deflectable positives*. Science 76:238
Anderson CD (1933) *The positive electron*. Phys Rev 43(6):491
Anderson CD (1983) *Unraveling the particle content of the cosmic rays*. The birth of Particle Physics. Cambridge University Press, p 131
Blackett P, Occhialini G (1933) *Some photographs of the tracks of penetrating radiation*. Proc Roy Soc A 139:699
Cowan E (1982) *The Picture that was not reversed*. Eng Sci 46(2):6
Dirac PAM (1928) *The quantum theory of the electron*. Proc Roy Soc A 117:610

Dirac PAM (1928) *The quantum theory of the electron. Part II*. Proc Roy Soc A 118:351

Dirac PAM (1930): *A Theory of electrons and protons*. Proc Roy Soc A 126:360

Dirac PAM (1931) *Quantised singularities in the electromagnetic field*. Proc Roy Soc A 133:60

Golden RL et al (1979) *Evidence for the existence of cosmic-ray antiprotons*. Phys Rev Lett 43:1196

Oppenheimer JR (1930) *On the theory of electrons and protons*. Phys Rev 35:562

Positron Photograph (Dec 19, 1931) *Cosmic rays disrupt atomic hearts*. Sci News Lett 20(558):387

Schuster A (1898) *Potential matter. A holiday dream*. Nature 58:367

Skobeltzyn D (1983) *The early stage of cosmic ray particle research*. The birth of Particle Physics. Cambridge University Press, p 111

Articles Related to Chapter 5

Aguilar M et al (AMS Collaboration) (2016) *Antiproton flux, antiproton-to-proton flux ratio, and properties of elementary particle fluxes in primary cosmic rays measured with the alpha magnetic spectrometer on the International Space Station*. Phys Rev Lett 117:091103

Aguilar M et al (AMS collaboration) (2019) *Towards understanding the origin of cosmic-ray positrons*. Phys Rev Lett 122:041102

Arqueros Martínez F (2009) *Las partículas más energéticas de la naturaleza*. Revista Vida Científica, nº 2:66

Bazilevskaya GA (2014) *Skobeltsyn and the early years of cosmic particle physics in the Soviet Union*, Astropart Phys 53:61

Cholis I, Linden T, Hooper D (2019) *A Robust Excess in the cosmic-ray antiproton spectrum: implications for annihilating dark matter*. Phys Rev D 99:103026

Choutko VA (2018) *AMS heavy antimatter*. AMS days at La Palma, La Palma (Canary Islands, Spain), 9–13 April 2018

di Mauro M, Manconi S, Donato F (2019) *Detection of a γ -ray halo around Geminga with the Fermi-LAT data and implications for the positron flux*. Phys Rev D 100:123015

Hess VF (1940) *The discovery of cosmic radiation*. Thought: Fordham Univ Quart XV:225

Linsley J (1963) *Evidence for a primary cosmic-ray particle with energy 10^{20} eV*. Phys Rev Lett 10:146

Logachev YI et al (2013) *Cosmic ray investigation in the stratosphere and space: results from instruments on russian satellites and balloons*. Adv Astron 2013, Article ID 461717

Poulin V, Salati P, Cholis I, Kamionkowski M, Silk J (2019) *Where do the AMS-02 anti-helium events come from?* Phys Rev D 99:023016

Skobelzyn D (1927) *Die Intensitätsverteilung in dem Spektrum der γ -Strahlen von Ra C*, Zeit Phys 43:354

Skobeltzyn D (1934) *Positive electron tracks.* Nature 133:23

Skobeltzyn D (1983) *The early stage of cosmic ray particle research.* The birth of Particle Physics. Cambridge University Press, p 111

Wulf T (1910) *Observations on the radiation of high penetration power on the Eiffel tower.* Physik Zeit 11:811

Articles Related to Chapter 7

Bambi C, Dolgov AD (2007) *Antimatter in the Milky Way.* Nucl Phys B 784:132

Blinnikov SI, Dolgov AD, Postnov KA (2015) *Antimatter and antistars in the Universe and in the Galaxy.* Phys Rev D 92:023516

Blinnikov SI, Dolgov AD (2014) *Stars and black holes from the very early universe.* Phys Rev D 89:021301

Cohen AG, De Rújula A, Glashow SL (1998) *A matter-antimatter universe?* Astrophys J 495:539

Cohen AG, De Rújula A (1998) *Scars on the CBR?* Astrophys J 496L:63

Dolgov AD, Silk J (1993) *Baryon isocurvature fluctuations at small scales and barionic dark matter.* Phys Rev D 47:4244

Dolgov AD (2016) *Early formed astrophysical objects and cosmological antimatter.* Int J Mod Phys A 31:28

Esposito S, Recami E, van der Merwe A (2009) *Ettore Majorana: Unpublished research notes on theoretical physics.* Springer

Fargion D, Khlopov M (2003) *Antimatter bounds from antiasteroid annihilation in collisions with planets and Sun.* Astro Phys 19:441

Fukugita M, Yanagida T (1986) *Baryogenesis without Grand Unification.* Phys Lett B 174:45

Gavela MB, Hernández P, Orloff J, Pene O (1994) *Standard Model CP violation and baryon asymmetry.* Mod Phys Lett A 9:795, e-Print: hep-ph/9312215

Khlopov MY (1998) *An antimatter globular cluster in our galaxy: a probe for the origin of matter.* Gravitation Cosmol 4:69

Khlopov MYu, Rubin SG, Sakharov AS (2000) *Possible origin of antimatter regions in the baryon dominated Universe.* Phys Rev D 62:083505

Majorana E (1937) *Teoria simmetrica dell'elettrone e del positrone* (A symmetric theory of electrons and positrons). Il Nuovo Cimento 14(4):171

Poulin V, Salati P, Cholis I, Kamionkowski M, Silk J (2019) *Where do the AMS-02 anti-helium events come from?* Phys Rev D 99:023016

Rehm JB, Jedamzik K (2001) *Limits on cosmic matter-antimatter domains from Big Bang nucleosynthesis.* Phys Rev D 63:043509

Sakharov AD (1967) *Violation of CP invariance, C asymmetry, and baryon asymmetry of the universe.* J Exp Th Phys Lett 5:24

't Hooft G (1976) *Symmetry breaking through Bell-Jackiw anomalies*. Phys Rev Lett 37:8

Zichichi A (2006) *Ettore Majorana, genius and mystery*. CERN Courier July 2006. https://cerncourier.com/a/ettore-majorana-genius-and-mystery

Articles and Books Related to Appendix B

Close F (2009) *Antimatter*. Oxford University Press

Fraser G (2000) *Antimatter. The ultimate mirror*. Cambridge University Press

Gato Rivera B (2018) *Antimateria*. CSIC/Los libros de la Catarata (Spanish)

Hanson NR (1961) *Discovering the Positron (I)*. Br J Philos Sci 12(47):194 (Published by Oxford University Press)

Hanson NR (1963) *The concept of the positron. A philosophical analysis*. Cambridge University Press

Positron Photograph (Dec 19, 1931) *Cosmic rays disrupt atomic hearts*. Sci News Lett 20(558):387

Schuster A (1898) *Potential matter. A holiday dream*. Nature 58:367

Skobeltzyn D (1983) *The early stage of cosmic ray particle research*. The birth of Particle Physics. Cambridge University Press, p 111

Wilson D (1983) *Rutherford simple genius*. Hodder and Stoughton

Printed in the United States
by Baker & Taylor Publisher Services